The Complete Guide to
Using Google in Libraries

The Complete Guide to Using Google in Libraries

Instruction, Administration, and Staff Productivity

Volume 1

Edited by Carol Smallwood

ROWMAN & LITTLEFIELD
Lanham • Boulder • New York • London

Published by Rowman & Littlefield
A wholly owned subsidiary of The Rowman & Littlefield Publishing Group, Inc.
4501 Forbes Boulevard, Suite 200, Lanham, Maryland 20706
www.rowman.com

Unit A, Whitacre Mews, 26-34 Stannary Street, London SE11 4AB

British Library Cataloguing in Publication Information Available

Library of Congress Cataloging-in-Publication Data Available

ISBN 978-1-4422-4689-8 (hardcover: alk. paper)
ISBN 978-1-4422-4690-4 (paperback: alk. paper)
ISBN 978-1-4422-4691-1 (ebook)

♾️™ The paper used in this publication meets the minimum requirements of American National Standard for Information Sciences Permanence of Paper for Printed Library Materials, ANSI/NISO Z39.48-1992.

Printed in the United States of America

Contents

Part V: Library Productivity

Foreword

As I approach my fiftieth birthday and twenty-seventh year in the library and information field, I find it easy to reflect on computers in the lives of librarians. While most librarians, especially younger ones, rightfully tune out when people start talking about Moore's Law and the rapid changes we have witnessed over the past few years, many aspects of these "old-timer" stories are applicable to librarianship today.

In my very first library job in the Archives and Manuscripts Department at the Chicago Historical Society (1988–1992), we created WordPerfect documents for archival finding aids and saved them to 5.25-inch floppy disks. Once I finished an archival finding aid or similar guide and it was printed up and available in our binders, I would delete the files. We had a forward-thinking consultant in one day who called this practice insane. It was just a few years later when I fully understood where he was coming from. While the cost of floppy disks was not huge, it is hard to imagine why we needed an electronic version of these documents when they were simply going to be printed and put in binders.

Less than ten years later, while at the National Baseball Library at the Baseball Hall of Fame and Museum, I was able to create a free public catalog using downloadable software (something increasingly new in the late 1990s) and MARC files that we had from our retrospective conversion of the book catalog at the library. While not perfect, it was usable and, more important, free. This experiment served as the basis for one of my favorite articles, "MacGyver's OPAC: Using Free Software and a Few Tricks to Create an OPAC."

In those few years, the entire scope of computing for libraries, especially smaller ones, changed dramatically. The young archivist who was deleting files became the person who figured out a way to make material more widely available. It wasn't a situation where we were blessed with more money or people, but more options and opportunities. The creativity we can apply to the tools we have at hand is ultimately the greatest advantage that we have in any technological environment. We need creative librarians who do not simply copy or replicate what other libraries have purchased, but find solutions that can be easily and freely implemented to solve new problems.

This brings me to this wonderful book. Just as I waxed poetic about the free OPAC I created back in 1997, librarians today are doing amazing things with tools that can be used for no other cost than our ability to learn how they work. Carol Smallwood's *The Complete Guide to Using Google in Libraries: Instruction, Administration, and Staff Productivity, Volume 1* provides librarians (and others) with creative and typically no-cost solutions to problems. Google's suite of tools, from Google Docs (now Google Drive), Google Scholar, Hangouts, Forms, and so on, are freely available to the Internet community. This book is divided into sections, each focusing on selections from Google's suite of tools that can be exploited in the library. The section headings are "Library Instruction for Users," "Collaborations," "Library Administration," "Collection Management," and "Library Productivity." Each section describes how librarians can take advantage of the suite of Google tools to change the way their library works without the burden of an additional bill. As you embark on this work, you will find ways to support library instruction, guiding researchers through Google Scholar as it works for your institution, using Google Hangouts, creating assessment tools, managing crucial documents (including license agreements), among many other tasks. The sections are made up of thirty chapters where you can see examples of how libraries have taken advantage of Google tools to improve their library operations.

This book will provide readers with examples of creative thinking that will allow libraries to do great things with a modest budget. Ideally, the creativity found in these pages will provide readers with ideas on how they might tackle problems that have not yet arisen. Google tools can help us now and well into the future as we design library programs. While there are thirty solutions here, my hope is that this is the creative inspiration for thirty thousand more.

Corey Seeman, director, Kresge Library, University of Michigan, and manager, *A Library Writer's Blog*

Preface

Google has vast resources that are often insufficiently used by librarians or library users. To remedy this, I sought chapters for this anthology by practicing public, academic, school, and special librarians as well as LIS faculty in the United States and Canada for creative, practical approaches that use Google applications. The call for chapters included the suggested topic "Google as a library partner or competitor" to encourage wide coverage; there was no contact with Google in connection with this anthology.

Volume 1 of *The Complete Guide to Using Google in Libraries: Instruction, Administration, and Staff Productivity* contains thirty chapters divided into sections on "Library Instruction for Users," "Collaborations," "Library Administration," "Collection Management," and "Library Productivity." The concise, how-to chapters are geared toward helping public, school, academic, and special librarians; library consultants; LIS faculty/students; and technology professionals. It was a pleasure working with these contributors, all of whom are willing to share their knowledge on this important topic with others in the field.

Acknowledgments

- Brady Clemens, director of Juniata County Library, Mifflintown, Pennsylvania
- Jane Devine, chief librarian of LaGuardia Community College, Long Island City, New York
- Karen Harrison Dyck, past president of the Canadian Library Association
- Lauren Magnuson, systems and emerging technologies librarian at California State University, Northridge
- Amanda McKay, director of Helen Matthes Library, Effingham, Illinois
- Sharon M. Miller, director emeritus at Mechanics Institute Library, San Francisco, and member of the Library Technology Industry Advisory Committee for San Francisco City College
- Vanessa Neblett, assistant manager at reference central of Orange County (Florida) Library System
- Lura Sanborn, media librarian at St. Paul's School, Concord, New Hampshire

Part I

Library Instruction for Users

Chapter One

App It Up

Using Google Apps in Library Instruction

Heather Beirne, Ashley Cole, and Sarah Richardson

Creating an active learning environment in library instruction can be challenging. When covering the ins and outs of researching in one-shot sessions, it can be tempting to simply lecture to students, offering "canned" searches of databases. However, lecture is becoming an increasingly outdated mode of instruction, and a new approach is needed to meet the needs of "digital natives," who tend to learn best when active in the classroom and also when their demand for innovative, technology-based learning is met (Magnuson 2013). One way to remain relevant and scaffold students' prior knowledge is to integrate Google and its apps into library instruction to address the larger digital, information, visual, and technological literacy picture.

Google Applications provide numerous tools that librarians can incorporate to promote active learning and enhance library instruction sessions. Many apps are versatile enough to use in a wide range of lesson plans for both face-to-face and online settings. In addition, they promote digital literacy while providing a highly collaborative learning environment. These apps help facilitate brainstorming activities and group work, allow for quick and easy feedback, and aid in planning and assessment. Using Google Apps in library instruction is an excellent way to enhance lesson plans and help strengthen students' technology skills and awareness. This chapter will briefly outline some basic functions and uses for various Google Apps, as well as provide ideas and inspiration for use of these tools in library instruction.

GOOGLE DRIVE AND PRODUCTIVITY SUITE

Drive

Drive is a cloud-storage service that includes a free productivity suite containing Docs, Slides, Spreadsheets, and Forms. Google's productivity apps, which are similar to some Microsoft Office applications, provide flexible, convenient ways for librarians to prepare instruction sessions, as well as an ideal working environment for students to develop their information literacy, digital literacy, and research skills.

These are some features and advantages of Drive and its associated apps:

- Drive storage space can be accessed from any computer, mobile device, tablet, and so on, with an Internet connection. Because all Google applications are based in the cloud, they are the same anywhere you use them, and accessing and editing your work from any computer or device is simple.
- Construct collaborative, active learning exercises in an online or face-to-face environment.
- Create, share, and collaborate on multiple documents, spreadsheets, presentations, and forms in real time.
- Access and download files in various formats, including Microsoft Office, Adobe, and more.
- Add collaborators and control viewing, editing, and commenting rights using the share feature.
- Customize sharing so that users are not required to sign in to access or edit files.
- The chat feature in each app encourages teamwork by allowing students to critically discuss any suggestions or changes.
- The comment feature enhances asynchronous work by allowing users to add notes for later review and collaboration.
- Add-ons are available to enhance apps, such as MindMeister, Letter Feed Workflows, and EasyBib. These can be found in the add-on store or directly from the add-on menu in Docs and Spreadsheets.

These features are helpful during all planning stages and delivery of library instruction, and when collaborating with colleagues on lesson plans and assessment (Evans 2014). Furthermore, as our students become more digitally savvy, Google Drive and its productivity apps can be used to engage and teach our students by immersing them in a learning process that allows everyone to explore, create, and share information.

Docs

Docs, a basic word processing app, allows users to create and format documents, translate them into another language, and view their revision history. Outlines and assignments can be shared to allow students to collaborate virtually with team members on a single document that everyone can edit in real time. This is more efficient than printing paper handouts or having students struggle with locating correct links during activities like website evaluation. Docs allows librarians to push links or share prompts with students, guiding them through online or face-to-face class activities, and also providing a space to add comments and complete exercises. Librarians and students can discuss answers that appear in a document, giving students the added incentive of having their work appear in the real-time document and projected onto the board for their classmates to read, creating an "authentic audience" for their work and motivating them to contribute meaningfully.

Mind mapping is one of the best ways to brainstorm, but it is sometimes deemed too daunting for an online environment. MindMeister, a Google Docs add-on, turns any bulleted list into a mind map. The first level of your bulleted list becomes the root of the map, while the rest branch out into second-level topics. This add-on is a great way to bring mind mapping into digital instruction. Other add-ons such as EasyBib, for example, can also contribute significantly to instruction. EasyBib allows students to create a bibliography without ever leaving Docs. Citations can be automatically generated from a title or URL and formatted into MLA, APA, or Chicago. EasyBib encourages users to be responsible in their research and, in turn, provides the basis for instruction on plagiarism and the importance of acknowledging sources. Letter Feed Workflows is another add-on of note. For librarians and instructors who provide instruction online, Letter Feed Workflows allows students to route their assignments for approval. Librarians may find this helpful as a tool for source checking, as students can route citations in Docs to the librarian to double-check for credibility before they actually begin writing their paper.

Slides

Slides provides presentation capabilities similar to Microsoft PowerPoint. Users can also easily recycle portions of previously created presentations by importing them into Slides. Group projects and presentation assignments can be collaboratively created and edited via the web, allowing students a creative outlet to show off their work virtually. In the library instruction classroom, for example, students can explore the library, databases, and collections, then create mini-presentations to demonstrate their knowledge and teach others online or face-to-face.

Forms and Spreadsheets

Spreadsheets allows users to create, update, and modify spreadsheets, similar to Microsoft Excel. Forms is useful for creating quizzes, questionnaires, and surveys for both active learning exercises and assessment purposes. When a form is created, Google automatically creates a spreadsheet for participants' responses, which are then compiled and available for immediate feedback, analysis, and assessment (Koury and Jardine 2013; Simpson 2012). Forms and Spreadsheets are ideal for creating pre- and post-tests for the purpose of evaluating student learning and performance. Whether instruction is online or face-to-face, forms can be shared with students in class, shared with the class instructor, or easily embedded into a learning management system. Forms may be used to support in-class activities, either face-to-face or on-line, as students can answer questions and do exercises alone or with a group. Students can further engage in active learning by collaboratively creating their own forms to quiz one another about the topic. Further, since Forms now allows for the inclusion of YouTube videos, it is perfect for creating multimedia-rich pre-class materials, which can be used in the flipped class-room, a model of instruction that involves pushing course materials such as readings, videos, or websites to students before class so that class time can be dedicated to activities that promote higher learning (Educause 2012; Davies, Dean, and Ball 2013).

ADDITIONAL GOOGLE APPS

Hangouts

Hangouts, an app within Google's social media site, Google+, provides a way to facilitate virtual group discussions. Hangouts provides an online meeting place for libraries that do not have access to a paid web-conferenc-ing platform. A hangout, or group conversation without video, can accom-modate up to one hundred participants. Users can also participate in video calls with up to ten participants (fifteen with a Google Education account). Google Hangouts potentially can be used in almost any classroom environ-ment—even massive open online courses (Bruff 2013). Librarians can use Hangouts to promote more engaging conversations, allowing students to share images, videos, and chat across multiple devices as if they were having a face-to-face conversation. The class can also break up into smaller work groups to stimulate self-directed learning and teamwork. Working together in this way empowers students to pool their ideas and skills to conquer complex issues, building their confidence (Levi 2014). Please note that Hangouts requires a large bandwidth to be fully functional. Librarians will also need to coordinate with faculty members in advance to get student e-mail addresses.

Images

Images is a platform to search the web for image content. Using Images in lesson plans for a variety of subject areas, from marketing and advertising to history and art, encourages students to critically evaluate and interpret images in context. The importance of these skills is outlined in the Association of College and Research Libraries' (ACRL) Visual Literacy Competency Standards (ACRL 2011). Google's Advanced Image Search tools also make it easy to locate results based on an image's usage rights. Image search results can be easily refined using a variety of criteria such as size, shape, color, type, date uploaded, and licensing information. For lessons on copyright and intellectual property issues, it is possible to "drag and drop" an image that you already have access to into a Google Image search in order to investigate its origins and copyright restrictions. When creating slides or docs for an assignment, students can use the skills learned from this lesson to responsibly use images in a presentation and engage in digital citizenship.

Moderator

Moderator is a crowdsourcing service in which users can submit comments and questions and also vote on existing questions and comments. Using Moderator, librarians can create a series or discussion forum about a selected topic and open it up for discussion to get instant feedback. Groups of students can submit questions, ideas, suggestions, and even YouTube responses to be answered by the administrator or other students. Using Moderator during lecture or as a free online discussion board encourages student interaction and provides a less-intimidating way for students to ask questions of the librarian, either during or after class.

Scholar

Scholar is a Google web search engine that allows users to search for scholarly content such as articles, books, government and legal documents, patents, and theses. Users are able to search for scholarly content across disciplines. Libraries using an OpenURL-compatible link resolver can enhance Scholar results for their patrons by displaying links to full-text resources available through online resources to which the library subscribes. Links to works that cite a selected article, as well as related articles, are also available. Scholar allows users to generate citations, export results to a citation manager, save results in their own personal Scholar library account, or set up alerts for new information related to a particular topic. For those libraries that do not subscribe to proprietary databases, Scholar can be used to teach search skills such as Boolean operators, truncation, and wildcards. Further, Scholar helps promote critical thinking in that it presents links to related articles and information such as "cited by" references. Scholar serves as a bridge between the Google generation and traditional library research by enabling students to use a familiar tool to access vital academic resources otherwise buried in the mechanics of library databases. However, Scholar occasionally retrieves nonscholarly content, which can be frustrating and confusing for users. Although Scholar retrieves a wide variety of content from many sources, it is imperfect at sorting content compared to library databases.

Sites

Sites makes creating a simple website as easy as editing a document. For libraries that do not subscribe to content management systems such as Lib-Guides, Sites is a great way for librarians to collect, organize, and share customized content for individual classes (Koury and Jardine 2013). Sites encourages librarians and students to brainstorm and share ideas in one place and to build content in a variety of formats. Creating a site allows users to move beyond a simple document and organize and embed multimedia and links as well as content from other Google applications. Creating pages within a site allows librarians to share content, such as videos and forms that are organized and easily navigable. A librarian may design their own site for use in a one-shot session, either online or face-to-face, in which each part or step within an assignment lives on a separate page to guide and assist the student through their coursework. Sites can also be used in place of course management systems that instructors use to introduce content. Librarians embedded within these sites can post links to relevant library resources as well as videos, quizzes, and other nonlibrary resources that pertain to topics covered by the course.

Voice

Voice gives users a free number that is linked to a personal Google account. It allows librarians to give out a Google Voice number to students (and other library users), giving them the ability to text or leave voice messages, which are automatically recorded and transcribed. This could be used with or in place of traditional e-mail, which may be of particular use for connecting with online students, as it creates the opportunity for a more personalized, voice-based interaction than e-mail allows. It is also possible to link other phone numbers to Voice, which can act as a manager for the numbers while keeping personal phone numbers private. Messages can also be downloaded and embedded and used in conjunction with other tools, such as Sites (Behmer 2013).

YouTube

YouTube is Google's popular video-sharing service. Videos can be easily embedded in other Google apps, such as Sites and Slides. By pushing a YouTube video in advance and flipping the classroom, librarians can free up valuable class time for hands-on, practical activities that promote information literacy. Librarians can even use real-life advertisements found on YouTube to promote students' critical thinking, making the work more purposeful. By combining this activity with Scholar, students can find resources to support or refute the claim made by the ad. Librarians can have students use YouTube videos to practice evaluation standards, such as accuracy and bias, an easy way to incorporate visual literacy. YouTube also offers education-related channels, including YouTube Teachers, which contain thousands of educational videos ideal for students at any level. Librarians can also use YouTube to upload tutorials and instructional videos for simple sharing. However, be aware that YouTube videos may disappear without warning, and advertisements, comments, and "related" videos are potentially distracting in the classroom.

CONCLUSION

The functionality of Google Apps is multifaceted, making the possibilities for enriching library instruction endless and easily applicable. Since Google Apps are free and in the cloud, no software has to be installed or updated on computers, documents are automatically saved, and multiple users can collaborate simultaneously. To participate in most app-facilitated class activities, such as those utilizing Forms, Docs, or Sites, students do not even need to set up a Google account. Further, the share feature gives students the option to return after class to review their work from any computer. Students can submit work digitally, creating a real-time, authentic, paperless body of work that can be saved, organized, and shared for assessment purposes. Librarians can also review student work created in class and see the history of revisions that were made to the document, providing a snapshot of students' thought processes in completing the exercise. Indirectly, using Google Apps in instruction provides a valuable introduction to this increasingly ubiquitous suite of applications, adding a digital literacy component and familiarizing students with a set of tools that they will likely use again and again in their academic and professional lives. Apps also accommodate different learning styles as students jump back and forth between technology and discussion. Class becomes more fast-paced, keeping all students engaged and on their toes.

However, there are several drawbacks for librarians, instructors, and students. Lessons using apps often require multitasking on the part of the instructor and students as they move back and forth between databases and apps. In addition, some students may need extra assistance when learning how to use a new app. In these situations, consider pairing them with another student who is more technologically savvy. Overall, however, many students may find Google Apps to be more user friendly than other apps or programs they have encountered. Additional planning will also be needed, since Sites, Docs, Forms, and other apps need to be prepared and possibly shared before class, such as in a flipped classroom setup. Though apps give librarians the flexibility to create highly customizable content, preparing and delivering an app-heavy lesson plan can be time consuming. Also, technology is unpredictable, and servers can go down during crucial times. Though individual users are able to configure offline access with Google Drive, allowing access to some files during server or Internet outages, the ability to collaborate with other users and work on files in real time is lost in these situations. Outages have the potential to temporarily disrupt active learning exercises and group work in the classroom. As with any lesson plan that uses technology, it is always advisable to have a low-tech backup plan to turn to in an emergency. Finally, features of Google Apps are constantly being changed and improved. Teaching innovatively with them requires a librarian to constantly stay informed as the apps are revised. One way to stay in the know is to follow the *Google Drive Blog*, which is consistently updated by the Google Drive Team with news and general ideas for use of Google Drive apps.

With all things considered, the advantages of using Google Apps in instruction far outweigh the shortcomings. These tools help support classroom work during the planning, implementation, and assessment stages. The applications covered in this chapter can be used individually or in many different combinations to enrich students' learning experiences both online and in the classroom, in any learning environment or academic discipline.

REFERENCES

ACRL. 2000. "ACRL Information Literacy Competency Standards for Higher Education." Last modified January 2000. Accessed March 20, 2014. http://www.ala.org/acrl/standards/informationliteracycompetency.

———. 2011. "ACRL Visual Literacy Competency Standards for Higher Education." Last modified October 2011. Accessed March 20, 2014. http://www.ala.org/acrl/standards/visualliteracy.

Behmer, Stacy. 2013. "Can You Hear Me Now? Google Voice and Other Ways for Speech to Text and Text to Speech with Google." Accessed March 20, 2014. https://sites.google.com/site/mwgs2013/can-you-hear-me-now.

Bruff, Derek. 2013. "Lessons Learned from Vanderbilt's First MOOCs." *Vanderbilt University Center for Teaching Blog*. August 19. Accessed March 20, 2014. http://cft.vanderbilt.edu/2013/08/lessons-learned-from-vanderbilts-first-moocs/ .

Davies, Randall S., Douglas L. Dean, and Nick Ball. 2013. "Flipping the Classroom and Instructional Technology Integration in a College-Level Information Systems Spreadsheet Course." *Educational Technology Research and Development* 61 (4): 563–80.

Educause. 2012. "7 Things You Should Know about Flipped Classrooms." Accessed March 20, 2014. https://net.educause.edu/ir/library/pdf/ELI7081.pdf .

Evans, Becky. 2014. "Google Apps in Classrooms and Schools: 40 Ways to Start Using Google Apps." Accessed March 20, 2014. https://docs.google.com/presentation/d/18YII85JZ8lXcKH-SvaXG7g0pW7vr8efZZsx5nL2Qimc/edit#slide=id.i0 .

Google Drive Blog. n.d. http://googledrive.blogspot.com/ .

Koury, Regina, and Spencer J. Jardine. 2013. "Library Instruction in a Cloud: Perspectives from the Trenches." *OCLC Systems & Services* 29 (3): 161–69. doi:10.1108/OCLC-01-2013-000.

Levi, Daniel. 2014. *Group Dynamics for Teams*. 4th ed. Thousand Oaks, CA: SAGE.

Magnuson, Marta L. 2013. "Web 2.0 and Information Literacy Instruction: Aligning Technology with ACRL Standards." *Journal of Academic Librarianship* 39 (3): 244–51.

Simpson, Shannon R. 2012. "Google Spreadsheets and Real-Time Assessment." *College & Research Libraries News* 73 (9): 528–49.

Springshare. 2013. "LibGuides by Springshare." Accessed April 7, 2014. http://www.springshare.com/libguides/ .

Chapter Two

Developing Research Skills with Developmental Students

Theresa Beaulieu

The developmental students described in this article are enrolled in a non-credit class, which is required. The students come to the library for three fifty-minute sessions, during which they are instructed on using the library-provided discovery tool, Google Scholar, the library catalog, and RefWorks, among others, so they can complete a research paper for the class. The paper requires students to investigate a career that they are interested in pursuing and to use scholarly and popular resources for their research. The assignment's guidelines direct students to use a variety of sources, articles, books, and a personal interview.

Academic librarians provide instruction for many different groups of students. Yet occasionally librarians seem less than interested in working with developmental students. There can be many reasons for this, but one that rankles is the deficit-model perceptions of developmental students, as it leads to low-level and irrelevant instruction sessions.

This perception is not uncommon in the larger culture. Such students can be perceived by instructors and others as having a deficiency and may be labeled "at-risk" or remedial students. These negative labels can encourage instructors to see such students as having deficits, instead of recognizing and building on the students' strengths. These negative perceptions can also lower the instructors' expectations of the students and can cause instructors to frame their instruction with a goal of "catching the student up" or cause them to try to fill in everything a student may have missed. In academia and popular media, the metaphor of the education gap and of student deficit is used widely, but it does not benefit the developmental students who experience such labeling.

In reflecting on these realities and the students enrolled in this class, I considered the following time-honored objectives to be essential when creating instruction for developmental students.

User-centered instruction. With a user-centered lens, developmental students are simply starting from a different place. In that regard, they are the same as honor students, who are also starting from a different place than the mythical "average" undergraduate student. When designing instruction for this group, consider what the students know and how they will connect with the instruction. In this case, I had experience with developmental students and had collected feedback from them about their experiences in the library. This review showed that developmental students do have a varied library skill set. Some do not have experience with libraries and assume that academic libraries charge for books, much like Redbox charges for video rentals. Other students are unfamiliar with things such as HTML, PDF, or e-mailing themselves a document. Still others understand all of this and want to get the job done with little interference from a librarian.

Focus on essential skills. Library skills that are considered basic should be evaluated to see if they actually are essential and if they are being taught at the right stage for the student. Some librarians seem to intuit this information and realize that while a doctoral student may need to locate and use datasets, on average a developmental student probably does not. On the other hand, unless one looks closely at the actual skills being taught, developmental students could waste their time learning about the differences between "Dewey Decimal" and "Library of Congress." Finally, technology makes some skills unnecessary or obsolete. Most students today can successfully research without knowing how to use an abstract or read microfiche. Many of the students in these courses have significant obligations outside of class, so being explicit about how to use these tools for research and for general use increases their appeal to students.

Build on strengths. Effective pedagogy begins instructing at the point of need and builds on what students already know. The developmental students in these classes are first-year students and are new to the university. They are familiar with some of the services that Google provides, and though they might not be able to articulate it, they are aware of its basic functions and its rhetorical structure. Engaging with students at their point of need and from their experience forms a natural connection and transition to scholarly research. For instance, an instructor can relate narrowing search results to limiting search results when using the retail website Zappos, and students understand the relationship immediately.

With these goals and the students' varied skill sets in mind, Google and the services it provides fit the bill. These services are adaptable to many different users and skill sets, they build on what students already know, and the skills transfer to other settings. It is my goal to ensure the instruction is user-centered and useful for the students. An added bonus of this approach is that it increases students' intrinsic motivation in the sessions, as they can see how the skills really benefit them in and out of school.

The following Google resources not only meet the above goals for working with developmental students, they help students progress academically. These tools and methods have been successfully used in the classroom and have received positive feedback from instructors and students.

CHROME

When students come to the instruction session and log on to the computers, they use Chrome as their Internet browser. Besides providing a common ground for comparing search results, Chrome has many features that facilitate searching for information. Other browsers have similar features, but it is helpful for everyone to have the same functionality in class.

Search Pages and Documents for Key Terms

Example: **control f**

One of the most useful functions that many students are not aware of is "control f." This function opens a search box that allows the user to search the page for specific terms. This is a great time saver when looking for a particular word or phrase in a document. A student looking for information about a nursing career may wonder if the article "Career Development: Graduate Nurse Views" has any information about nurses' satisfaction with their salaries. If the "control f" function returns no hits on the word *salary*, the student can move on without reading or scanning the whole article, looking for information that is not there.

GOOGLE

Students sometimes approach the research assignment with baggage from instructors in other classes or from high school, where they have been told such things as "Don't use the Internet" and "You can't just Google it." In fact, they *have to* use the Internet, and they *can* Google it. This negative perception of the Internet and of Google needs to be turned around. Instead of forbidding students from using these tools, show students how to use them effectively for academic purposes. One of the best things about Google is that students are familiar with it. Instructors can build on that familiarity.

Some of the best Google features that can assist developmental students are as follows:

Check Spellings

When demonstrating a search, an instructor can purposely misspell a word, then show students how to find the correct spelling with Google, which can generally figure out what you are looking for. For example, a Google search for "goverment job," will display results for "government job." The same search for "goverment job" in our local discovery tool results in zero hits, and a student may come to the conclusion that the library does not have any information on the topic. Advising students to use Google to check their spelling if they are not getting the necessary search results will help them succeed.

Clarify What You Are Looking For

Since Google uses more natural language than most databases, it is easier to find what you are looking for without the exact words that many other databases require. One student in the class wanted to find information on a medical career. He wanted to be the person that takes the x-rays. Rather than having me feed him the words *x-ray technician* or *radiologist*, the student Googled "who takes x-rays." That search led to sites with the names of several occupations that deal with x-rays and explained the differences between the various jobs. As it turned out, the student wanted to be a radiologic technologist—a term I could not have come up with, as I had never heard of it.

Find Specific Information on a Cluttered Internet Site

Example: **site:www.dol.gov nursing career**

Students can find information about their careers on regular Internet sites, such as the Department of Labor, but the sheer quantity of information available can be baffling, and students may struggle to find what they are looking for. Using Google site searching allows students to use Google's search engine on a specific site when using Chrome as their browser. This function can be demonstrated by going to the Department of Labor's Internet site and asking students where on the page they can find information about a nursing career. Usually this results in false starts or a limited amount of information. An instructor can then demonstrate by copying the URL of the home page of the Department of Labor and pasting it into the Google search box (prefaced with "site:") and add the keywords "nursing career," which results in web pages within the Department of Labor site with information about a nursing career (see example above).

Time the Students' Work

Example: **timer 5 minutes**

Instructors often give students time in class to search for information about their topic. Using Google's timer function allows both the instructor and the students to keep track of time just by glancing at the screen in the classroom. To use this function, type "timer 5 minutes" in the Google search box. It will begin counting down and will beep when time is up.

GOOGLE SCHOLAR

While students are quite familiar with Google, they have seldom encountered Google Scholar, although it is an essential tool for academic research. It can be helpful to explain that Google Scholar is a search engine, not a database, and why that matters. This can be done by explaining that Google Scholar searches across many databases, so it will find results that students have access to via their library's online subscriptions, results that are available because of open access, and results that students cannot access without utilizing interlibrary loan. It does not bother students that they cannot immediately access all the articles once they know the reason why. Another benefit of using Google Scholar is that all the search techniques used in the regular Google search engine will work in Google Scholar, so students are able to build on recently acquired skills. Here are some more reasons students find Google Scholar to be useful.

Refine Search Terms

Once students find an article that relates to what they are looking for, they can reuse those terms in subsequent searches. One former student provided a useful tip: have students use their phones to photograph a successful search in order to replicate it later.

Limit the Age of the Results

Information about working in the field of nursing from twenty years ago is not helpful for a student who wants to become a nurse today, so students should use the date limits to customize their results accordingly. Here is one place where it is useful to point out facets, which is a feature similar to that found on many commercial sites. Facets show up to the left of the search results.

Find More Resources

Google Scholar increases the number of results returned compared with our local discovery tool. In order to explain why this is so, reiterate the difference between a search engine and a database. While some critics see Google Scholar's high rate of return as a negative feature, students do not. They are not going to search through all four million results. They will limit the results and use only the ones they can access from the first few pages, or they will review their terms and try again.

Find Open Access Articles

In addition to returning more results, Google Scholar also returns more open access articles. Discuss why this is the case, and inform your students about government-funded research and the open access movement.

Find a Scholarly Article You Have Found Once Before

Often students have done some research on their own before coming for library instruction. Once they learn about RefWorks in class, they want an easy way to find the same article again so they can send it to RefWorks. If they know the article's title, students will be able to retrieve that article by typing the title into Google Scholar. Students can also put the title in quotes, although this requires careful spelling. While most libraries have a citation linker, it is often less user friendly than Google Scholar for such a search.

Find Other Resources That Are Related to the Article

Under each of the results Google Scholar returns are links to "cited by" and "related articles." Related articles can be especially useful for students in this class.

Cite the Reference Material

This can be done in one of two ways in Google Scholar:

1. Because these library sessions include RefWorks instruction, students learn how to import references to RefWorks using Google Scholar. To activate this feature, click on the settings gear at the top of the Google Scholar home page. This will open the Scholar settings page. The default page is the "Search results" page, which is indicated in the left-hand column. At the bottom of the page is "Bibliography manager." Choose the radio button in front of "Show links to import citations into" and then use the arrows to select RefWorks. Save the settings.

This will generate "Import into RefWorks" links under each search result.

2. Students also are shown how to create citations one at a time with Google Scholar, as Scholar can be easier to use than RefWorks for short papers. To create a citation with Google Scholar, locate the row of links under each search result. Then click on the "More" link. Currently, this will display the "Cite" link, which will display citation information for the article in various citation styles.

Use Google Scholar from Home

Because many students in this class have responsibilities outside of school, they are often doing research from off-campus locations. Setting their Google Scholar preferences to show links to their library will identify which results they have access to via library subscriptions. In order to do this, click on the settings gear at the top of the Google Scholar home page. This will open the Scholar settings page. Select "Library links" from the left column. Enter the name of the students' library in the search box and search. The results will display below the search box. Place a checkmark in the box before the name of the students' library and the "Open WorldCat" box that appears. Save those choices. Now when searching in Google Scholar, links to the library's resources (if available) will appear to the right of each result. Click on these links and not the title to view the resource. The title generally takes you to the publisher's website.

Find Only PDFs

Example: **filetype:PDF**

Since Google Scholar returns many articles for this type of search, students may find it useful to limit their results to materials to which they will have immediate access.

Advanced Search

Google Scholar also has an advanced search; however, this is not often demonstrated to developmental students. This feature can be very helpful when an individual student needs to narrow information or if the student has an unusual topic. The advanced search can be accessed by clicking on the down arrow in the search box.

GOOGLE BOOKS

Since Google Scholar returns results for scholarly articles, citations, and books, we will now turn to Google Books. Not all students are familiar with Google Books, but they are curious about its potential use for this paper.

Save Time by Finding the Right Book before Going to the Stacks

In addition to books showing up in a Google Scholar search, many library catalogs display links to Google Books in "limited preview" (when available) so students can see if a particular book is what they are looking for before traversing the stacks.

Increase the Number of Books Available on the Topic

Going directly to Google Books increases the number of books available for research. Sometimes students only need a small amount of information from a book. If the information they need is readily available from Google Books, it saves the student time in finding a physical copy of the book.

Realize That It Is Okay to Read a Portion of a Book for a Research Project

Students are already aware that the whole book most likely is not available in Google Books. They are generally not aware that it is accepted practice to read and use portions of books for their research.

YOUTUBE

Students are generally adept at using YouTube. However, they may not have considered its academic purposes.

Explain Concepts to the Class

YouTube videos are used in these classes to explain the difference between popular and scholarly articles. One benefit is that librarians do not have to make their own videos to explain concepts to the class. Another benefit is that the use of such videos breaks the potential monotony of one speaker. The links to the videos are easily embedded in research guides.

Legitimize Scholarly Uses of YouTube

By using YouTube videos in the classroom, students gain a greater apprecia-tion of YouTube's usefulness for academic work. This gives them implicit permission to use such videos to teach themselves what they have identified as necessary to learn.

* * *

All told, Google is an excellent tool for introducing students to academic research. One thing to remember is that Google makes changes to its tools incrementally, rather than all at once as some databases are wont to do. One recent change allows students to create their own library of research, which is a common feature of academic databases. Make sure you keep current with Google in order to eliminate confusion when demonstrating its functions to students. An easy way to do this is to subscribe to a Google blog. You can just Google that; it is on the Internet.

REFERENCE

Cleary, Michelle, Jan Horsfall, Paulpandi Muthulakshmi, Brenda Happell, and Glenn E. Hunt. 2013. "Career Development: Graduate Nurse Views." *Journal of Clinical Nursing* 22 (17–18): 2605–13.

Chapter Three

Enhancing Access to Master's Thesis Research with Google Fusion and Google Maps

George L. Wrenn

INTRODUCTION

This chapter describes how the Humboldt State University (HSU) Library in Arcata, California, is using Google applications to enhance access to master's theses archived on Humboldt Digital Scholar, HSU's institutional repository (Wrenn 2014a, 2014b, 2014c). HSU grants master's degrees in a number of programs that emphasize field or community-based research. In the College of Natural Resources and Sciences, graduate students in master's programs conduct research in Humboldt County, greater Northern California, and beyond; in the social sciences, students often do research in community-based settings. This chapter will show how to visually represent thesis research locations using Google Spreadsheets, Google Fusion, Google Maps, and supplemental applications.

PROJECT SCOPE AND CONTEXT

The cataloging librarian at HSU initiated and leads the project with support from a library staff member and works on the project intermittently as time permits. Our goal is to map all theses with a field or community-based research focus. Approximately one-third of a collection of 875 theses has been completed since the project began in early 2012; approximately 200 more are in process, with additions planned as new master's theses are added to the repository. The project began with four programs with a field-research focus: biology, fisheries, forestry, and wildlife. Future maps will also present these additional programs: the environment and community, environmental resource engineering, environmental systems, geology, natural resources, natural resources planning and interpretation, social work, sociology, wastewater utilization, wastewater management, watershed management, and wildlife management.

OTHER RESEARCH MAP PROJECTS

A number of colleges and universities have visualized faculty and student research, including:

- Clemson University Graduate Program in Historic Preservation, which presents student thesis research using Google Earth (Clemson School of Planning, Development, and Preservation 2014)
- New York University Center for the Study of Human Origins, which presents a static map of student research locations (New York University Center for the Study of Human Origins 2014)
- Purdue University Department of Anthropology, which presents a map of faculty research (Purdue University, Department of Anthropology 2014)
- Texas A&M, College of Geosciences, Department of Geography, which presents a student research map (Texas A&M University, College of Geosciences, Department of Geography 2014)
- UCLA Fielding School of Public Health, which presents a Google map of faculty and student research, training, and service activities (UCLA Fielding School of Public Health 2014)
- University of Washington Information School, which presents a Google research map (University of Washington Information School 2014)

As this list demonstrates, many projects focus on current research or a specific program. Although we plan to plot the research locations of all the theses in our collection, maps may be presented successfully on a smaller scale.

GOOGLE AND OTHER APPLICATIONS USED IN THE PROJECT

When the project began in early 2012, we selected the Library of Congress's Viewshare program as a tool for creating interactive maps. Viewshare, which is free to registered users, can provide "distinct interactive visual interfaces to your digital collections, including maps and timelines, and sophisticated faceted navigation" (Library of Congress 2014). Although Viewshare is easy to use, with a flexible and appealing set of presentation options, we decided to experiment with maps in Google Fusion because Viewshare maps lack

Google's terrain and satellite views, which may provide additional context for understanding a research area. If, for example, a forestry thesis examines the after effects of a forest fire, Google's satellite view may reveal the burn area. Google also has the advantage of familiarity. Google claims that "1 million active websites" use the Google Maps API, reaching "1 billion people every week" (Google 2014e). Google is also appealing because HSU supports a number of Google applications, including Google Drive, which has proven very useful for sharing project documents, and Google Spreadsheets, which supports real-time collaborative editing, streamlining data preparation.

Google Fusion, introduced in 2009 and still in development, has the ability to create thesis maps. Fusion is a "service for sharing and visualizing data online. It allows you to upload data, share and mark up your data with collaborators, merge data from multiple tables, and create visualizations like charts and maps" (Langen, Madhavan, and Shapley 2009). Unfortunately, HSU does not support sharing Google Fusion data with non-HSU users: a university login is required for access (Google's support documentation states that "your domain administrator or service provider can turn off particular services or restrict your ability to move data to or from the organizational account" [Google 2014b]). As a workaround, we created a personal Google account, which makes it possible for anyone to access the Fusion map. We continue to create maps with Viewshare because we find it user friendly, with useful search and sort functions. (Because this chapter's focus is Google, Viewshare is minimally discussed.)

WORKING WITH SPREADSHEET FIELDS

We use DSpace software to manage our institutional repository collections (DSpace 2014). Twenty-five metadata fields based on Dublin Core (DCMI 2014) are used in thesis description:

dc.contributor	dc.description.abstract	dc.language.iso
dc.contributor.advisor	dc.description.program	dc.publisher
dc.contributor.author	dc.description.provenance	dc.relation.ispartofseries
dc.contributor.other	dc.description.sponsorship	dc.subject
dc.date.accessioned	dc.format.extent	dc.subject.lcsh
dc.date.available	dc.format.mimetype	dc.subject.other
dc.date.issued	dc.identifier.other	dc.title
dc.description	dc.identifier.uri	dc.title.alternative

dc.type

We export all thesis metadata into a spreadsheet, where it is readied for use in Google Fusion. Export provides an excellent opportunity to identify and correct errors that have crept into the metadata over time. In our data, inconsistent application of language codes during the submission process created four copies of each column. Although this duplication quadrupled the size of the spreadsheet, clean-up was straightforward (if slightly time-consuming): we merged related columns into one, without loss of data, using the spreadsheet's merge command. We also sorted columns to discover empty cells and supplied missing data as necessary.

To prepare spreadsheet data for presentation, column labels should be put into a form suitable for presentation (with words spelled out and consistent capitalization) and unwanted columns removed. To avoid overloading the map user with thesis metadata, we decided to retain only seven of the twenty-five columns: Title, Author, Advisor, Program, Issue Date, Description, and the URI (Uniform Resource Identifier) link to the repository record. The Description column contains the type of data found in a MARC record's dissertation note field: degree, institution, and date. Advisor and Program become filters on the map display so that the user may select, for example, all biology theses or all forestry theses with a particular advisor. We omitted columns considered of less interest to users, including the abstract (too lengthy to display), publisher (always "HSU"), subjects (contains uncontrolled, user-submitted keywords), and a number of repository-related columns.

ADDING MAP-RELATED FIELDS TO THE SPREADSHEET

With thesis-related columns in place, map-related columns must be added to the spreadsheet. Some of this data will display on the map; some is for internal use. Data to be displayed includes:

- Study Site/Area
- Study State/Region
- Study Country
- Coordinates
- Notes

The Study State/Region, Study Country, and Notes columns are not needed to generate the map. State and country data are provided to support filtering on that data. The Notes column specifies where to go in a thesis for information on the research area (e.g., "study area, p. 6; map, p. 7; coordinates given:

41°45'37"N, 124°15'W). Four nonpublic columns are then added to help organize geographic data:

- Study Site (yes/no): *indicates the presence/absence of field research*
- Mapping Note: *information about coordinates assigned to a thesis*
- Embargo (yes/no): *identifies theses that will not be mapped due to access restrictions*
- Icon: *to assign a different map pointer to each program*

The Mapping Note provides a way to document where and how coordinates were assigned. The Icon column is used to assign a different colored map pointer to each program, following naming conventions used in Google Fusion (e.g., "large_blue," "large_red," "large_green") (Google 2014d).

ASSIGNING STUDY SITES

With geographic columns in place, the most labor-intensive aspect of the project may begin: recording the locations of field research. We proceed program by program, beginning with programs that emphasize field research. Each thesis must be examined closely to determine whether it contains field research and, if it does, where the research was conducted. In many theses, a "study site" chapter is present. Information from this chapter (or elsewhere if the chapter is not present) is recorded in the spreadsheet columns Study Site/Area, Study State/Region, and Study Country.

It is important to enter data consistently, especially state and country data, so that it may be sorted in Fusion. For example, if research took place in California, the Study State/Region column should not contain variations of California such as "CA" or "Calif." Data in the Study Site/Area column does not lend itself to sorting because we usually transcribe what is written in the thesis and make no attempt to control naming conventions. Some examples:

- Along the Eel River and several ocean-fronting sandy beaches within Humboldt County
- Boyes Creek, a tributary of Prairie Creek, located in Redwood National and State Parks, Humboldt County
- Cosumnes River Preserve near Galt
- Grand Teton National Park and the National Elk Refuge, and northern range of Yellowstone National Park
- L. W. Schatz Demonstration Tree Farm, located near Maple Creek
- Olympic Peninsula, Clallam County
- Southeastern portion of the Trinity Alps Wilderness, Klamath Mountains

- Yuba and Nevada County foothills in the central portion of the Sierra Nevada Mountain Range
- Trinity River Hatchery

As these examples demonstrate, study sites vary greatly in extent. In some theses, a single location is identified with precise geographic coordinates; in others, several distinct geographic areas are studied. One of the more problematic theses encompasses dozens of localities from Canada to California:

> Collection localities were distributed from the Skeena River in British Columbia, Canada, south to the Ventura River, California. Multiple collections were made in larger drainages including the Skeena, Fraser, Columbia, Umpqua, Coquille, Rogue, Klamath, Sacramento/San Joaquin and San Francisco Bay drainages. A diverse range of stream habitats, flow regimes, topography, and climate conditions are encompassed within the collections distributed over approximately 2600 km of the Pacific coastline. (Goodman 2006, 5)

In this case, we used text from the thesis abstract to designate a study area: "2600 km of the North American Pacific coast from the Skeena River, British Columbia to the Ventura River, California" (Goodman 2006, iii).

Some theses include research that may not be suited for plotting on the map. If a biology student has examined fossils from collections around California, should the entire state be considered a research area? Some research is completely lab based. Should the lab be plotted on the map? HSU theses in education, psychology, and sociology have raised the most questions: a number of theses survey residents of a populated area, such as a town or county. Should that location be included on the map? Although we have not established rules for problematic theses, we carefully document decisions and refer to them as new ambiguities arise.

ASSIGNING GEOGRAPHIC COORDINATES

After study sites have been added to the spreadsheet, the corresponding geographic coordinates must be assigned. These are entered in a single spreadsheet column in decimal form (e.g., 40.507453, -121.434288). Google Maps offers a quick way to produce decimal coordinates: simply click on a location whenever the arrow cursor is visible. iTouchMap is one of many online applications that may be used to visualize and assign coordinates (iTouchMap.com 2014). It indicates latitude and longitude for any map point selected, displaying results in a pop-up window for copying. For example, selecting a spot near Kneeland, California, produces the following coordinates:

Latitude–Longitude:

40.71465, -124.017792
Lat: 40°42'52.7394"
Long: -124°1'4.0512"

iTouchMap also converts from the degree-minute-second (DMS) format to the decimal format required by Google Fusion and vice versa.

For theses with extensive study areas or multiple study sites, a map point can only approximate the location(s) of research and give only a general sense of the research area(s). For demographic research at the city, county, or state level, placing coordinates at a government building may have to suffice. Map pointer placement often becomes a matter of judgment. What point should be used for a thesis with thirteen named study sites "within and around Redwood National and State Parks, California" (Sun 2012, 16)? If most of the research is clustered in one area, a point entered in that area will provide a general sense of research locations. Thesis metadata and the thesis itself may be consulted for clarification. When multiple scattered study sites exist, selecting one may be necessary. For example, another thesis states that "bats were captured and echolocation calls were recorded at locations across the eastern United States, ranging north to central Indiana, south to Louisiana, east to Pennsylvania and North Carolina, and west to Missouri" (Corcoran 2007, 7). In this case, our only recourse was to identify a point somewhat central to those five states (we selected a spot in central Indiana [39.504041, -86.132812]). We use the Mapping Note to document choices. (One such note reads: "Coordinates: Lone Mtn, Oregon, 1 of 16 study sites. Map p.10, study sites p.11," indicating that one of sixteen study sites was chosen.) We also note whether coordinates were obtained from the online Geographic Names Information System (GNIS), described below, or another online place-name database.

FINDING PLACE NAMES

Various online tools may be used to look up coordinates by place name. PlaceNames.com provides a searchable database of U.S. place names and their corresponding latitudes and longitudes (PlaceNames.com 2014). The United States Board on Geographic Names has created standard nomenclature for U.S. geographic names, accessible through the GNIS database (U.S. Geological Survey 2014a). This database is very useful for locating coordinates for "feature names" in various feature "classes." For example, Eureka, California, is classified as a "populated place" with the designation "county seat" and coordinates (40.8020712, -124.1636729) corresponding to the location of the United States Post Office and Court House in Eureka (U.S. Geological Survey 2013). Yosemite National Park is classed as a park located in four counties; coordinates are provided for thirty-four topographical features within the park (U.S. Geological Survey 2014b). Coordinates for non-U.S. locations may be looked up in the NGA GEOnet Names Server, a site maintained by the U.S. military; map locations link to Google Maps and MapQuest (National Geospatial-Intelligence Agency 2014).

USING AN AUTOMATED SCRIPT TO GEOCODE

An alternative to manual assignment and online lookup of coordinates is the use of a script that automatically generates coordinates in the spreadsheet. We use a script that was installed as an add-on to our Google Docs spreadsheet. This script, "Geo for Google Docs," is available on GitHub.com (GitHub.com 2014). It examines data in our Study Site/Area field to find geocodes using three online map applications: OpenStreetMap, Yahoo PlaceFinder, and Cicero. Installation was simple: Google Spreadsheet includes a Script Editor (accessible on the Tools menu) into which source code may be pasted. After naming and saving the script, a new spreadsheet menu tab, named Geo, became available.

For the script to function, three new columns must be added to the spreadsheet next to the Study Site/Area column: Geo_latitude, Geo_longitude, and Geo_accuracy. The script is run by highlighting the Study Site/Area column and selecting "Geo Addresses" on the Geo menu tab. If successful, the new columns are populated with data. The Geo_accuracy column identifies place name features. Common features for our data include administrative, bay, hamlet, natural reserve, park, residential, river, and water. This data helps with coordinate assignment, but we do not present it to users.

Although the script is designed to streamline assignment of coordinates, we have found that significant editing of Study Site/Area column data is necessary for the script to work effectively. Data in this column is often transcribed verbatim from thesis text in a form not understood by OpenStreetMap, Yahoo PlaceFinder, or Cicero. When we experimented with more concise forms of naming in a parallel column, results were better, but the script is still not successful in many cases. We continue to rely on manual geocoding.

Comparing coordinates provided by the script, GNIS, and Google Maps, it became clear that each uses a different system for assigning coordinates. For example, the script put coordinates for the Shasta River (41.5512461623583, -122.397410542507) near the town of Weed (Siskiyou County, California). The GNIS database coordinates are a few miles north (41.8306956, -122.5941924), where the Shasta meets the Klamath River. A Google Map search places the coordinates (41.771078, -122.590834) between the two, somewhat closer to the GNIS coordinates. When choosing among different coordinates, we generally favor GNIS because it represents an established, "official" source of information.

FINAL STEPS: CONVERTING THE SPREADSHEET TO A FUSION TABLE

To prepare data for presentation in Google Fusion, a "presentation copy" of the master spreadsheet is created. Within this copy, rows without coordinates are deleted (representing theses not yet mapped or not being mapped), and all but the following columns are eliminated:

Advisor	Study Site/Area
Author	Study State/Region
Title	Study Country
Issue Date	Notes
Program	Icon
Description	Coordinates
URI	

Google Drive includes a "connect more apps" function that makes Fusion accessible to Google Drive documents. The spreadsheet may then be converted to a Fusion table (Google 2014a). Google Fusion offers many of the same features as a spreadsheet; map data could, at this point, be entered and maintained entirely in Fusion. A Google help topic, "Fusion Tables or a Spreadsheet?," outlines the benefits of each application (Google 2014c).

After the spreadsheet is converted, a title (e.g., "Map of Field Research Sites in Selected Humboldt State University Master's Theses") and descriptive information (e.g., "A frequently updated map of field research from selected Humboldt State master's theses in biology, fisheries, forestry, wildlife, and wildlife management") are provided. The Geocode option in the map menu is used to display map pointers based on data in the Coordinates column. After publishing the table (by changing visibility from private to public in the Tools menu), it can be shared as a link or embedded on a web page (Wrenn 2014a). Metadata is presented in tabbed views, to which filters may be applied. The defaults are a map view, a spreadsheet view (rows of data), and a card view that organizes data vertically, for example:

Advisor: Camann, Michael A.
Author: Beeler, Heather
Issue Date: 2009–2012
Program: Biology

Description: Thesis (M.A.)—Humboldt State University, Biological Sciences, 2009

Title: Community Succession in Macroalgal Wrack Implications for Prey Resources of Breeding Western Snowy Plovers (*Charadrius alexandrinus nivosus*) on Northern California Beaches

URI: http://hdl.handle.net/2148/556

Study Site/Area: Clam Beach/Little River State Beach

Notes: Study area, p. 9; Coords: 10406218E, 4538743N

Study State/Region: California

Study Country: USA

Coordinates: 40.987508, -124.118386

MAP POINTERS IN VIEWSHARE AND FUSION

We sometimes assign the same coordinates to multiple theses because students do research in the same area. Viewshare's map pointers display a number when more than one thesis is associated with a point; we have configured these pointers to show a list of titles when clicked on. Unfortunately, Google Fusion lacks this feature. If multiple theses share the same map point, only one of them displays when the pointer is selected; the others are invisible. Reassigning coordinates may be necessary in such cases.

MAPS IN GOOGLE EARTH

We are also experimenting with Google Earth, which offers an attractive, rotatable, 3D view of Earth. Google Earth uses Keyhole Markup Language (.kml) files; several steps are required to convert a spreadsheet file to .kml format. The Earth Point website provides a number of tools to assist with conversion and summarizes the necessary steps (Earth Point 2014). We also use Google's Spreadsheet Mapper 3 tool, which provides a "starter spreadsheet" and a detailed walkthrough of the conversion process (Google 2014f). Although we do not intend to take advantage of Google Earth's more advanced options, we are intrigued by the many ways that map layers may be applied to enhance the visual experience of the user.

CONCLUSION

Using Google applications, online tools such as iTouchMap, and a geocoding script, we have developed maps that provide a visually informative introduction to research in HSU's master's programs. As Google's Big Picture Group states, "information visualization can make complex data accessible, useful, and even fun" (Cichowlas et al. 2014). We hope our maps are useful to those exploring the wealth of academic field research conducted in Humboldt County and beyond and that the work of HSU's graduate students is now more accessible to students, faculty, and the larger community.

REFERENCES

Cichowlas, Alison, Colin McMillen, Jon Orwant, Fernanda Viégas, Martin Wattenberg, and Jim Wilson. 2014. Big Picture Group. https://research.google.com/bigpicture .

Clemson School of Planning, Development, and Preservation. 2014. "Theses." Accessed June 9, 2014. http://www.clemson.edu/caah/pdp/historic-preservation/students/theses.html .

Corcoran, Aaron J. 2007. "Automated Acoustic Identification of Nine Bat Species of the Eastern United States." Master's thesis, Humboldt State University. http://hdl.handle.net/2148/169 .

DCMI (Dublin Core Metadata Initiative). 2014. "DCMI Metadata Terms." http://dublincore.org/documents/dcmi-terms/ .

DSpace. 2014. http://www.dspace.org/ .

Earth Point. 2014. http://www.earthpoint.us/ .

GitHub.com. 2014. "Geo for Google Docs." https://github.com/mapbox/geo-googledocs .

Goodman, Damon. 2006. "Evidence for High Levels of Gene Flow among Populations of a Widely Distributed Anadromous Lamprey *Entosphenus tridentatus* (Petromyzontidae)." Master's thesis, Humboldt State University. http://hdl.handle.net/2148/121 .

Google. 2014a. "Create: A Map." https://support.google.com/fusiontables/answer/2527132?hl=en .

———. 2014b. "Data Access by Your Administrator or Service Provider." Privacy and Security. https://support.google.com/accounts/answer/181692?hl=en .

———. 2014c. "Fusion Tables or a Spreadsheet?" https://support.google.com/fusiontables/answer/171191?hl=en&ref_topic=2592746 .

———. 2014d. "Map Marker or Icon Names." https://www.google.com/fusiontables/DataSource?dsrcid=308519#map:id=3 .

———. 2014e. "Our History in Depth." Google Company. Accessed June 9, 2014. http://www.google.com/about/company/history/ .

———. 2014f. "Spreadsheet Mapper 3." Google Earth Outreach. http://www.google.com/earth/outreach/tutorials/spreadsheet3.html .

iTouchMap.com. 2014. "Latitude and Longitude of a Point." http://itouchmap.com/latlong.html .

Langen, Anno, Jayant Madhavan, and Rebecca Shapley. 2009. "Google Fusion Tables API." *Google Developers Blog.* Accessed June 9, 2014. http://googledevelopers.blogspot.com/2009/12/google-fusion-tables-api.html .

Library of Congress. 2014. Viewshare. Accessed June 23, 2014. http://viewshare.org/ .

National Geospatial-Intelligence Agency. 2014. NGA GEOnet Names Server. Accessed June 9, 2014. http://earth-info.nga.mil/gns/html/index.html .

New York University Center for the Study of Human Origins. 2014. "Student Research Locations." Accessed June 9. http://www.nyu.edu/gsas/dept/anthro/programs/csho/pmwiki.php/Home/StudentResearchLocations .

PlaceNames.com. 2014. http://www.placenames.com/ .

Purdue University, Department of Anthropology. 2014. "Purdue Anthropology Faculty Research." Google Maps. https://maps.google.com/maps/ms?msid=207647707939410347982.0004ad63d1fced7823e75&msa=0 .

Sun, Michael Cheng. 2012. "*Batrachochytrium dendrobatidis* Prevalence in Northern Red-Legged Frogs (*Rana aurora*)—10 Years Later." Master's thesis, Humboldt State University. http://hdl.handle.net/2148/1234 .

Texas A&M University, College of Geosciences, Department of Geography. 2014. "Student Research Map." http://geography.tamu.edu/prospective-students/student-research-map .

UCLA Fielding School of Public Health. 2014. "Global Projects." http://ph.ucla.edu/research/global-health-fsph/global-projects .

University of Washington Information School. 2014. "Research Map." https://ischool.uw.edu/research/map .

U.S. Geological Survey. 2013. "Feature Detail Report for: Eureka." Geographic Names Information System. Accessed July 7, 2014. http://geonames.usgs.gov/apex/f?p=gnispq:3:::NO::P3_FID:277605 .

———. 2014a. "Domestic Names." United States Board on Geographic Names. http://geonames.usgs.gov .

———. 2014b. "Feature Detail Report for: Yosemite National Park." Geographic Names Information System. Accessed June 10, 2014. http://geonames.usgs.gov/apex/f?p=gnispq:3:::NO::P3_FID:255923 .

Wrenn, George. 2014a. "Map of Field Research Sites in Selected Humboldt State University Master's Theses." Google Fusion. Accessed June 19, 2014. https://www.google.com/fusiontables/DataSource?docid=19xbuhSPgNbepQvjO50lV3Fzoa1mFWR2EyBMjMtwZ .

———. 2014b. "Research Study Sites Mapped for Selected Humboldt State University Master's Theses ." Viewshare. Accessed June 19, 2014. http://viewshare.org/share/965b70a4-f83e-11e3-a1f5-0ad109b0e86f/ .

———. 2014c. "View Maps of Biology, Fisheries, Forestry, Wildlife, and Wildlife Management Master's Thesis Field Research." Humboldt Digital Scholar. Accessed June 20, 2014. http://humboldt-dspace.calstate.edu/handle/10211.3/121711 .

Chapter Four

Google Trends as an Academic Research Tool

Expanding Scholarly Search Strategies

Sarah Cohn and Amy Handfield

Traditional academic library research focuses on subscription databases, with many librarians actively discouraging the use of the open Internet for academic research. However, subscription databases and resources do have limitations. With particular regard toward business, sociology, communication, and marketing students, most subscription databases do not provide the kind of market or culture research students need. A business student will need "Business Insights: Essentials" for detailed company data, or an overview of an industry, but in order to get real-time data about an industry or an area, the resource is lacking. A sociology student will use a subscription database for scholarly or news articles, but that tells her nothing about the current interest surrounding the issue. Google has several free analytic options that can provide these students with just the information they need; the most accessible and useful of these is Google Trends.

Google's increasingly robust data-collection mechanisms allow for interesting data to be taken from it. Google Trends is an easy introduction to the type of data Google collects and makes available to the public. Google Trends tracks searches done through Google from 2004 onward. Google Trends shows real-time data by tracking and displaying the relative volume of searches done on a term or topic. Since results can be filtered on a map, regional data is readily available. Some short-term future forecasting is included as well. While not a stand-alone research tool, Google Trends is a good complement to subscription resources.

For marketing and business students who do not have access to paid market data or research, Google Trends can offer a simple, free way to do market research. Google Trends can help students get information on industry, brand, product, or cultural trends by providing

- market and new market research,
- company research,
- competitor comparison, and
- industry keyword research.

This information can help students think about seasonality, advertising messages, brand associations, regional interests, and language. Students of sociology and other research-based social sciences will also find Google Trends to be of use, offering a glimpse into the collective mind.

HOW GOOGLE TRENDS WORKS

As with all things Google, Trends is intuitive to use. Google Trends allows for comparison of up to five different search terms. Enter your search term or terms, and the results are shown in an interest over time graph, a regional interest map, and a related searches list that shows search topics and search terms, or queries. If using more than one term, each is given a different color, for ease of comparison. Standard Google search strategies—quotation marks, + to indicate "or"—work in Google Trends.

The search boxes have two different search options: search term and search topic. Search term is your chosen natural-language term, and search topic is Google's suggested broad topic. As you enter your search term, Google Trends will suggest a list of topics that are closely related. These suggested topic terms may be more specific, but your results will be different from your original search, as the search topic function is still in the beta (final testing) phase. The search topic results include all different queries that may relate to the concept. The search term results include only searches that have that specific text. If the word *Manhattan* is used as the search term, the first suggestion for a topic is "Manhattan; New York City borough." A side-by-side comparison of the search term "Manhattan" and the topic "Manhattan; New York City borough" highlights these differences. The results for search term "Manhattan" will include any search done about the borough, the cocktail, the Manhattan Project, or any other use of the word. The results for the topic "Manhattan; New York City borough" show only searches relating to the specific borough of Manhattan. In some cases, the results are quite different, and in other cases, the results are practically identical. This is evident in comparing the search term "Obama" and the search topic "Barack Obama;

44th U.S. President." It is recommended users search with both options and examine the results for the search topic and search term. Users will find each option to have varying degrees of usefulness depending on their particular needs.

When teaching Google Trends in an instruction session, showing a basic search will give students an idea of how it works. An instructive search is to compare "heater" with "air conditioner." This search (figure 4.1) clearly illustrates what Google Trends results look like, and what information can be gained from it.

The interest over time graph shows the relative number of searches done on each term. Moving the mouse along the graph lines shows the month, year, and relative number of searches done. The numbers do not represent the total number of searches, but rather the volume of searches relative to the highest point on the chart. An option for news headlines allows you to easily see news stories related to your terms, displayed with letters along the graph

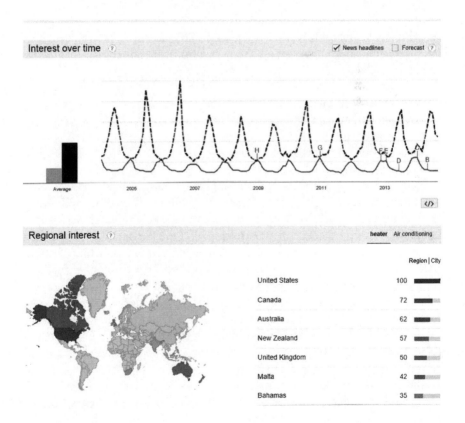

Figure 4.1.   Basic search using Google Trends.

lines. However, these are not always the most relevant headlines to your search. As usual with Google, it is unclear what algorithm they are using to determine what is included on the graph as a noteworthy news story. For a better news search, it is recommended to do a traditional Google search using the News option.

The regional interest map shows the search data from countries using Google to conduct searches. Data is not available for all parts of the world, but the interactive map will easily show where the data is coming from. The map allows for zooming in on an area by clicking. For example, clicking on the United States will break down the results by state, and further by county and city or town. Regional data is useful for students who are researching local issues that might not get coverage in larger studies. Regional data is also useful for those wishing to target a specific area with a marketing campaign by learning the region's preferred language surrounding an item or topic.

GOOGLE TRENDS AND ACADEMIC RESEARCH

As with any search, when starting a Google Trends search it is important to have research questions in mind. Some useful questions to ask of Google Trends include:

- What phrases or concepts are moving up?
- What phrases or concepts are in decline?
- What is and isn't popular?
- What are common variations of search terms?

A student creating a marketing campaign for a new natural food product would benefit from looking at the terms "organic food" and "natural food." Both terms are popular ways to denote healthfulness. The graph shows that both have had varied search popularities. Over time "organic food" has been a more popular search term that has fallen off somewhat, while "natural food" is gaining in search popularity. Using the map to look at regional preferences in the United States shows that "natural food" is a more prevalent search in Maine and "organic food" is more prevalent in Vermont. The campaign can be tailored to use language more common in specific geographic regions based on the data available in Google Trends. Looking at the related search topics to these two sets of terms is also instructive, as the top related search for "organic food" is "organic baby food." The top related search for "natural food" is "natural dog food." This shows a very clear distinction in people's minds about the difference between the terms "organic" and "natural."

A marketing or public relations student who needs to create a social media campaign plan might look at the following search terms: "Twitter," "Instagram," "Pinterest," and "Tumblr." Twitter has an overall larger search volume, but that number is in decline. Tumblr's search volume has leveled off, but Instagram's search volume is on the rise. Pinterest has the overall lowest search volume, and the number is fairly consistent. These trends would suggest to the student how best to use social media in a marketing campaign by showing which platforms are worth focusing on, and the most effective way to get the message out via social media platforms.

A sociology student researching the recent trajectory of the gay marriage movement could compare the terms "gay marriage" and "same sex marriage." The term "gay marriage" has a much higher overall search volume, so it could provide better data about overall and regional interest. A student studying perception of alternative energy sources would find the data from the search terms "solar energy" and "wind energy" useful, particularly the regional map. It shows very clearly which regions are searching the different types of energy.

Business students interested in company or competitor research are best served by using the search topic option for specific company names. A search for two relatively small sportswear companies, Under Armour and Lululemon Athletica, shows that there is a marked seasonal component to searches for these companies and within those smaller interest variations. A new sportswear company attempting to compete within the market would learn the best season to launch their company or new lines of clothing.

CONCLUSION

Google Trends offers a unique and specific view into the collective Internet mind of today's society and can be mined for useful and instructive data. Students of certain disciplines will find it more useful than others, specifically business, sociology, communication, and marketing students. However, even the general user will appreciate the data available and find something of interest in the results. Google Trends does have some shortcomings with its search functionality and resulting data. Like all Google products, there is no index of terms. Trends requires users to really know their search terms in advance, or work them out in a trial and error process. Terms that have more than one meaning can provide inaccurate results. It is also important to keep in mind that the results can be overly broad and provide only general trend data. Despite these drawbacks, Google Trends provides students with information that they do not generally have access to and can easily be incorporated into an academic research setting.

Chapter Five

Helping College Students Make the Most of Google Scholar

Aline Soules

As a faculty member teaching a two-credit information literacy course to first-year university students, I have observed that students frequently use Google in preference to any other tool. In spite of encouraging them to use library resources as a first rather than a last resort and in spite of teaching them techniques to do that, students go right back to Google. They say: "It's too difficult to use library resources. Google is easier." One response to this statement is, "At least use Google Scholar." Some first-year undergraduates have never heard of it. Those who have are often unclear about how it differs from Google, what it provides, or how to use it effectively. What they know is the word *Google*.

After several such exchanges, I explored Google Scholar in greater depth with the goal of persuading students that it is a better alternative than regular Google for scholarly work and the hope that, ultimately, some students might choose to use library resources. In fact, as students spend more time in university, their use of Google Scholar appears to increase. Shen (2012) conducted a quantitative study of graduate students' use of Google Scholar, identifying ease of use, access, and comprehensiveness as key factors. Moving to more experienced users, Chen (in Antell et al. 2013) points out that researchers use Google Scholar more often than most subscription-based abstracting and indexing services.

Google Scholar is definitely part of the information landscape, and its use increases with more research sophistication. Because students are familiar and comfortable with Google, they should be receptive to Google Scholar, and librarians should be able to use it as a bridge to library resources while helping students begin their use of Google Scholar with good techniques.

Google Scholar can also be used to instill information literacy fundamentals, an added bonus.

GOOGLE SCHOLAR

What is Google Scholar? Like Google itself, it is a search engine, in this case focused on scholarly materials in multiple disciplines, multiple formats, and from multiple traditional and nontraditional sources. Since launching in 2004, it continues to expand. The challenge is helping students understand its limitations and how to interpret their search results.

With library databases, information has traditionally been more contained and ordered. Google Scholar, however, is one big mix of formats (e.g., articles, books, abstracts, court opinions), publishers (academic, professional, trade, nontraditional), and open sources (online repositories, faculty websites, journals, etc.). To compete, library database vendors now offer "discovery services," which essentially replicate Google Scholar's approach, providing "one-stop searching" across library holdings and other sources, including Google Scholar, if the library chooses to include it. Asher, Duke, and Wilson (2012) compared the following: two discovery services, Google Scholar, and conventional library resources. They conducted a qualitative and quantitative study on students' searching habits. In all cases, students relied on default search settings and were unable to evaluate sources effectively. These authors suggested that focus on a single tool such as Google Scholar might allow librarians to concentrate on concepts and evaluation rather than explaining different ways of approaching different search systems.

This comparison underlines the need for information literacy education. Without necessarily intending to, the article offers a good reason for teaching students techniques in Google Scholar not only because they are comfortable with Google, but also because, after students graduate, most will no longer have access to library resources.

GOOGLE SCHOLAR FEATURES

- Students can search scholarly literature from a single search box (although there is an advanced search feature as well).
- Students can click on related works, citations, or live links to authors and other elements to take them to a different set of results.
- Students can click on the title for more information about that title—an abstract or, in some cases, full text, although there are challenges with this (see below).

- Students can set up alerts for e-mail notification of new Google Scholar items in a chosen search.
- Students can create an author profile. Although that sounds oriented more toward faculty who want to track their own publications, students need a profile to save searches or receive alerts.
- Students can check where an article has been cited after its publication, a service that currently is not available in many library resources. One exception is Thomson Reuters products such as humanities, social sciences, and science citation indexes.

Students are used to using Google (in whatever form) by simply typing words into the single search box and getting results, although of varying relevance depending on what they search. In introducing some of the above features to students, one challenge is convincing them that these techniques will make their research experience better rather than more complicated. The goal is not to make Google Scholar seem as complicated to them as library resources, but to focus on the idea that Google Scholar is better than regular Google for scholarly purposes.

SEARCHING

Students have trouble choosing search terms. Two Google features mask their inability in this regard. First, students often enter sentences or questions, for example, "what are the pros and cons of euthanasia?" Google ignores some words (called stop words), but words like *pro* and *con* give different results from words like *advantage* and *disadvantage*, although both retrieve results. This is an opportunity to help students understand the language of the academic world. A test of Google Scholar with the question "what are the pros and cons of euthanasia?" on December 21, 2013, retrieved "about 3,180" results. The first four were these (the matching term is in bold in Google, but given here after the title):

- Dying with Dignity: Understanding Euthanasia (matching term: euthanasia)
- Doctors' Opinions on Euthanasia, End of Life Care, and Doctor-Patient Communication: Telephone Survey in France (matching term: euthanasia)
- Hymenorrhaphy: What Happens behind the Gynaecologist's Closed Door? (matching term: what)
- Dying in Dignity: The Pros and Cons of Assisted Suicide (matching terms: pros cons)

The third unrelated result was retrieved by the word *what*. It can be difficult to convince students to enter a term or short phrase rather than sentences or questions, but their results will improve. A live demonstration can help persuade them, particularly if it is possible to put screens side-by-side to show different results depending on how terms are entered.

In addition to the use of terms, other search techniques are applicable in search engines and library resources generally. Examples include using quotation marks around a phrase to ensure that it is searched as a whole phrase; using the Boolean operators "or" and "not" ("and" is the default operator in Google); searching in site; and wildcards. A set of mini-tutorials was created to help students understand various techniques. These are listed in appendix 5.1 and available for use by any reader or student.

COVERAGE, ACCURACY, AND ORDER OF RESULTS

What are students searching? Google Scholar crawls the Internet and gathers bibliographic records from various sources, some of which also provide the same information to libraries, for example, JSTOR and Science Direct. Google Scholar also scoops up information from free resources, for example, PubMed, institutional repositories, open educational resources, and so on. Unlike some library resources that limit retrospective coverage, Google Scholar searches a full journal run as long as it is on the web. This increases results and can be good for certain subject searches. Google Scholar also covers nontraditional sources, for example, institutional repositories, as equally as traditional sources, often providing more extensive coverage of nontraditional sources at this point in time. On the other hand, Google picks up everything, published or not. For example, if a researcher mounts a working paper on his website, it will be picked up by Google Scholar along with published articles. This heightens the importance of evaluation because not-yet-published materials may or may not have undergone peer review or even editorial review. Also, if Google Scholar is gathering information from the very sources to which a library subscribes, why not use the sources themselves? This question can be raised with students later, once they are more adept with Google Scholar and more willing to consider library resources for in-depth and focused searching.

Google corrects no source errors in a record. Further, there are stories of errors created by Google's software programs that analyze language. Antell et al. (2013) refer to a time when Google's algorithms created phantom authors, for example, creating "P Login" out of "Please Login." Because the gathering process is machine-generated without human intervention, there is no human check on individual errors. As a result, not only are errors transported and perpetuated, new errors may be introduced to the system. On the other hand, Google works to improve its algorithms. P Login no longer exists, although a search for P Login provides results, the first three matches being RP Login, G Login, and P Devlin (P being the match in this last case, which is not particularly useful).

Further, Google has proprietary control over its programs. According to O'Kelly (Antell et al. 2013, 281), Google employees themselves are not entirely sure how Google Scholar works. O'Kelly notes that before Google Scholar's Metrics page was revised in April 2012, one statement read: "Since Google Scholar indexes articles from a large number of websites, we cannot always tell where (or if!) a particular article has been published." The statement may be gone, but is the issue resolved?

Google's goal is to rank documents "the way researchers do" (Google 2014). Google Scholar claims to weigh the full text of each document, where it was published, the document's author, and how often and how recently the document has been cited in other scholarly literature. For a document to be cited, however, it must be in the literature long enough for others to have included it in a bibliography for a newly published article. This can affect the currency of documents on the first screens. Well-cited documents are likely to appear earlier, while less-well-known or newer documents may appear on later screens, which students may not reach. To counteract this, it is possible to "sort by date" instead of "relevance" or to choose dates such as "any time," "since 2013," "since 2012," "since 2009," or select a "custom range."

It is also possible to "game" the system to affect rankings. López-Cozar, Delgado, Robinson-García, and Torres-Salinas (2012) created six documents by a "fake" author and uploaded them to a researcher's personal website, resulting in an increase of 774 citations in 129 papers.

These examples are good teaching moments, helping students to understand the importance of evaluation and the limits of machine algorithms. Many students want an "easy" answer to evaluation. They regularly voice the idea that if a website ends in .com, it's bad, but if it ends in .edu or .org or .gov, it's okay. Showing them studies about gaming the system helps them to think beyond a simple formula. Author qualifications, peer review (if that can be determined from a Google Scholar result), currency, relevance, context, and other considerations can be introduced.

This is also an opportunity to teach students the difference between keywords and subjects, particularly Library of Congress subject headings, which are created and assigned by human beings. Human intervention increases relevance and students pay attention to the idea that humans are smart while computers are dumb and do only what they are programmed to do.

RELATED WORKS, LINKS, AND CITATIONS

In any search process, students should be encouraged to think about follow-ing two lines of inquiry. One is to find information. The other is to look for clues—other terms not yet considered, other links, and other possibilities that might help them find new or better information. Each Google Scholar result is followed by these live links:

- Cited by [number]
- Related articles
- All [number] versions
- Cite
- Save
- More (appears occasionally)

"Cited by [number]" provides a screen with all the documents that have cited the original result. This is an opportunity to teach students about the nature of academic discourse. If an academic paper appears in the literature at a partic-ular point in time, the bibliography tells the reader what came before and allows the reader to follow the trail the author followed. The "cited by" list tells the reader what comes later: who read the paper and cited it in his or her new work. This helps students to understand that academic discourse is es-sentially a conversation in writing and that each document is a data point in the conversational stream.

"Related articles" gives a random number of similar articles, emphasizing the importance of looking for clues and treating the search process as a detective hunt. Of course, some clues lead to blind alleys, but some lead to more or better answers.

"All [number] versions" shows the document's sources. There is much duplication in the information world as publishers place their materials in as many abstracting and indexing services as possible. If the full text cannot be reached by clicking on the title in the results list, another version might lead to the full text. At this point, students need to learn more about how to connect to full text from this search engine, namely

- how to link Google Scholar to their libraries, and
- the format of the document they want. Is it a book? An article? Something else?

LINKING TO LIBRARIES

Linking Google Scholar to students' local libraries enables students to see what documents may be available to borrow, download, or print. This is easy, if they know how to click on "settings," then "library links." A search box lets students enter and save up to five local libraries, one at a time. This can include a mix of libraries or multiple names for the same library, for example, California State University, East Bay, and Cal State University, East Bay.

One library that is automatically connected is WorldCat, although that connection can be disabled. OCLC describes WorldCat as the "world's largest network of library content and services" and it contains records created by librarians and other information professionals. There is also a feature that allows students to enter a zip code to find the nearest library where a document may be found. Linking to WorldCat helps students to see the vast possibilities for retrieving a document. On the other hand, WorldCat can be frustrating if students cannot get the full text because they lack affiliation to the owning library or the library is too far away or if the item is unavailable through interlibrary loan, another service that needs explanation.

Linking to libraries, particularly libraries where students are affiliated, enables students to see the phrase established by their libraries to find full text. At Cal State East Bay (CSUEB), the phrase is "Find Text @ CSUEB." This connects students to CSUEB's holdings and shows them where online or in the physical library they can find the full text. Unfortunately, if their computer's IP address (Internet Protocol address, a numerical label unique to each device and identifying its affiliation) is not part of the library (in fact or by proxy), Google Scholar will often take students to a screen where they are asked to pay. Students may look for other documents in frustration, particularly if they keep running into "pay" screens. If they have time, they may call the library, in which case, they are told not to pay and given alternatives for finding the full text. In cases where students are desperately working at the last minute, however, they may pay and the library may or may not find out about it.

This is where looking at all the versions may help. One of these versions might take the student to the full text without payment. This is also an opportunity to teach students how to find a particular article in their local libraries and, if it is not available, how to use interlibrary loan. This gets their attention because free rather than fee makes them sit up and take notice. It also begins their process of connecting to library resources.

FORMAT

Google Scholar results are a mix of material types, which must be identified by students. Knowing the format of a result is helpful in finding it. If "Find Text" works, students may be taken to the logical place for a book (the library catalog) or an article (a particular database), but when that link does not work, knowing if something is a book or an article or a thesis or a conference paper helps students look in the appropriate place. This is another "bridge" to the library, and this is where understanding citations becomes important. If students can interpret a citation, they can look in the right place. A tutorial titled "Decoding Citations" was created to help students understand the "language" of citations, and it is listed in appendix 5.1 for reader and student use. Students may find citations frustrating and seemingly far removed from their lives, but if they can learn the language of texting, they can learn the language of citations. Once they can interpret a citation, they know whether to look in the library catalog, a database, or another source.

CITE, SAVE, AND MORE

Clicking on "Cite" causes a box to pop up with a citation for that result. Citations are currently given in MLA, APA, and Chicago formats (Turabian, not author-date). While these citations provide the bibliographic portion of the citation relatively accurately, students need to know enough to validate the citation, and they also need to add access elements, such as the database where it was found, a URL, a doi (digital object identifier, a character string that is unique to an individual document or object), the date of access, or whatever a specific format requires.

When students click on "save," the result is saved in "my library"; however, students must create a Google Scholar profile to save search results. This raises the issue of privacy, because students should be made aware that Google now has access to any information they enter in their profile. At a minimum, this will be their name and e-mail, although students may also add a list of "interests," which Google can use.

Clicking on "More" opens up links to "Library search" and, if students have linked to their libraries, a "Find Text" link.

ALERTS AND AUTHOR PROFILES

Setting an author profile not only "saves" results, it enables "alerts." Students can click "create alert" to bring up an "alert query," which is the search string students entered in the first place. Students can modify the search string at this point. Also on this screen are students' e-mails, the number of results to be shown per screen (ten or twenty), "update results," and "create alert." Once "create alert" is clicked, alerts are sent by e-mail when new documents are added to Google Scholar.

VIRTUES AND VICES

Google Scholar offers speed and efficiency and returns results. Coupled with its familiarity, it's no wonder students love it. The issue is how to help them to understand its limitations and how to use it in conjunction with other tools and resources. Unlike library resources that allow students to separate articles from book chapters from other types of materials, Google Scholar results are mingled, requiring interpretation. Further, the sources and order of results are not always clear. Are key relevant articles included? Are the results on the first screen the most current results? Do students take time to determine this? Students, particularly undergraduates, are often unaware and unconcerned. As long as they get something to use in their papers, they are satisfied.

Over time, Google Scholar has incorporated more features, but there are still fewer options than in library databases, although Google Scholar does provide the very useful "cited by" feature.

What draws students is Google Scholar's ease of use and guarantee of results, no matter what. It is the librarian's job to translate those results into meaningful understanding and help students learn that there are times when other resources may be better for a particular purpose.

CONCLUSION

Moving students from Google to Google Scholar provides a bridge between regular Google and library resources. It offers something of everything—ease of searching, a list of results, some sorting ability, basic citations in commonly used formats, links to other sources, an alert service, and familiarity. There are other sources that may provide more in-depth or focused results, greater ability to separate results into discrete formats, and slightly better citations in more formats, but students know Google and are less intimidated by Google Scholar than by library resources.

To take advantage of that bridge, here is a summary of some key teaching points to help students take better advantage of Google Scholar and ready them to use library resources as they gain more confidence:

- Search techniques: See appendix 5.1, tutorials.
- Coverage: See Google Scholar's "metrics," which lists the covered publications by discipline.

- Source: Discuss machine crawling and algorithms and how that can affect results.
- Accuracy: Discuss minor errors and larger content issues.
- Order of results: Discuss Google Scholar's criteria that affect the order of results and how many screens students should examine.
- Evaluation: Discuss various criteria in the context of Google Scholar and also in comparable library resources. Help students see the advantages and limitations of each.
- "Related" works and links: Explore how to combine a search for information with a search for clues to improve results.
- Citations: Discuss how to interpret them and how to use both bibliographies and "cited by" results.
- Format: Explain the importance of format and how citations can provide clues to understanding the format of documents for effective retrieval of full text.
- Save: Discuss "my library" and the implications for privacy.
- Alerts: Explain how to make information come to students rather than students repeating searches.

There are many teaching opportunities in Google Scholar, both for information literacy and for more effective use of the tool. All of this can be transposed to the library's resources when students have gained comfort and expertise with Google Scholar. The shift then will be less of a leap. A great exercise is to split a class into groups and assign to each group a different source—Google Scholar, relevant library databases, the catalog—and compare results.

Google Scholar is here to stay and improving all the time. It is useful and a great way to lead students into academic research.

APPENDIX 5.1

Tutorials

Effective Web Searching

- http://library.csueastbay.edu/modules/effective_web_searching/effective_web_searching.htm
- Covers the basics of good web searching
- Length: 3 min., 49 sec.

Web Tips (each of the following covers one search term)

- "Exactly," length: 1 min., 35 sec. http://library.csueastbay.edu/modules/web_tip_exactly/web_tip_exactly.htm
- "Not," length: 2 min., 19 sec. http://library.csueastbay.edu/modules/web_tip_not/web_tip_not.htm
- "Or," length: 2 min., 1 sec. http://library.csueastbay.edu/modules/web_tip_or/web_tip_or.htm
- "Quotation Marks," length: 2 min., 32 sec. http://library.csueastbay.edu/modules/web_tip_quotation_marks/quotation_marks.htm
- "Search in Site," length: 1 min., 57 sec. http://library.csueastbay.edu/modules/web_tip_search_in_site/search_in_site.htm
- "Terms," length: 2 min., 37 sec. http://library.csueastbay.edu/modules/web_tip_terms/web_tip_terms.htm
- "Wildcard," length: 2 min., 15 sec. http://library.csueastbay.edu/modules/web_tip_wildcard/web_tip_wildcard.htm

Decoding Citations

- http://library.csueastbay.edu/modules/decoding%20citation/decoding%20citation.htm
- Helps students learn how to interpret citations
- Length: 8–10 min., depending on how long it takes for users to click the correct answers

REFERENCES

Antell, Karen, Molly Strothmann, Xiaotian Chen, and Kevin O'Kelly. 2013. "Cross-Examining Google Scholar." *Reference & User Services Quarterly* 52 (4): 279–82. doi:10.5860/rusq.52n4.279.

Asher, Andrew D., Lynda M. Duke, and Suzanne Wilson. 2012. "Paths of Discovery: Comparing the Search Effectiveness of EBSCO Discovery Service, Summon, Google Scholar, and Conventional Library Resources." *College & Research Libraries* 74 (5): 464–88. http://crl.acrl.org/content/early/2012/05/07/crl-374 .

Google. 2014. "About Google Scholar." Google Scholar. Accessed October 17, 2014. http://scholar.google.com/intl/en-US/scholar/about.html .

López-Cózar, Emilio Delgado, Nicolás Robinson-García, and Daniel Torres-Salinas. 2012. "Manipulating Google Scholar Citations and Google Scholar Metrics: Simple, Easy and Tempting." *EC3 Working Papers* 6 (May 29). http://arxiv.org/abs/1212.0638 .

OCLC WorldCat. 2001–2013. http://www.worldcat.org .

Shen, Lisa. 2012. "Graduate Students Report Strong Acceptance and Loyal Usage of Google Scholar." *Evidence Based Library & Information Practice* 7 (4): 96–98. doi:10.1016/j.lisr.2011.02.001.

Chapter Six

Using Google for Primary Sources

Alejandro Marquez

The ability to look at primary sources and compare them to the original article allows the reader to verify and evaluate information. Google, Google Scholar, and Google Books (all three of which will be referred to as Google for simplicity) make finding these primary sources easier. The benefit of using the Google search engine as an information recovery tool is Google's ability to execute multiple searches retrieving many more results than a traditional library database. It can access nontraditional and archival sources that have links to full text, citations, or archival finding aids. However, not all resources and materials are available on the Internet. The use of Google along with traditional library resources such as subscription databases and archival materials will return better search results.

PRIMARY SOURCES VERSUS SECONDARY SOURCES

Primary sources offer the firsthand experience of an individual who participated in a historical event or time period or created materials in the course of their daily life, reflecting the viewpoint of the participant or author. Such sources include original documents such as autobiographies, memoirs, letters, interviews, official records, diaries, and oral histories. Sometimes the sources are a creative work such as poetry, art, drama, music, and novels; they can also be physical objects such as tablets, medals, pottery, furniture, clothing, and buildings. Primary sources are based on their content rather than their format, whether it be microfilm, a YouTube video, or a sound recording.

A secondary source interprets or analyzes primary sources. Secondary sources have the benefit of hindsight, offer analysis, and place events within a historical context. They offer readers a combination of the raw material of primary sources and interpretations by historians. Secondary sources can include popular books and magazine articles.

Tertiary resources provide overviews of topics by gathering and summarizing information from other resources. They provide the reader with a condensed version of the data that is easy to understand and interpret. Examples include encyclopedias, chronologies, almanacs, and textbooks. These resources are great for gathering background information on a topic but are usually not considered scholarly for higher levels of education.

However, the differentiation between primary and secondary sources is not always clear cut. The answer often depends on the topic, the academic discipline, and the context in which the sources are used by a researcher. For example, scholarly articles and manuscripts are a primary source for scholars in many academic disciplines including history, sociology, health sciences, and education. In computer science, a software program would be considered a primary source while the user's manual for the software would be a secondary source. Another example, this time from the field of culinary science, is a recipe and critique of that recipe being a primary and secondary source, respectively.

Additionally, a secondary source can be used as a primary source if a researcher is looking at the ways that information was used during a certain time period. An article published about the American civil rights movement in the *New York Times* can be read as a secondary source. It can also be seen as a primary source of how scholars interpreted the mores and attitudes during that particular time period.

TYPES OF SOURCES

- *Printed texts*—These materials can be both primary and secondary sources depending on the content. Examples include articles, pamphlets, letters, diaries, memoirs, autobiographies, and textbooks.
- *Serials*—Newspapers, periodicals, and magazines can be both primary and secondary sources depending on the academic discipline. The publication date of a serial can also be an important factor in the determining whether it functions as a primary or secondary source. As stated earlier, these serials offer a window into the attitudes and thoughts of scholars of a particular time period.
- *Maps*—These primary sources provide a cultural context. Maps are influenced by the political, social, and scientific events of their time and may contain inaccuracies, whether unintentional or deliberate. The Henricus Martellus world map reflects the latest scientific theory circa 1490 about the form of the world, portraying it on a flat surface. It is said that this map

convinced the rulers of Spain to allow Christopher Columbus to sail to the New World.

- *Government Documents*—Public records produced by a government are primary sources. They show the government's activities, functions, and policies. These sources include research reports, statistical publications, fact sheets, handbooks and manuals, presidential papers, budgets, maps and atlases, pamphlets, hearings (testimony), congressional reports, committee prints, records of proceedings and debates (Congressional Records), statutes, and bills.
- *Visual Material*—These primary sources reveal the subjective nature of how societies and individuals viewed themselves. These include visual, nontext, and nonaudio sources and can be one of a kind or reproduced. Visual material sources include photographs, films, paintings, drawings, prints, designs, and three-dimensional art such as sculpture and architecture.
- *Dissertations*—These materials are based on original research and created to fulfill the requirements for a doctoral degree. Usually considered secondary sources, the majority of their research resources is based on primary sources.
- *Audio*—Sound recordings, music, and oral histories reveal the daily habits and values of a society and are considered primary sources. Other related primary sources include musical scores, musical instruments, and other materials related to music and musicians.

SEARCH STRATEGIES USING GOOGLE

John Harvey Kellogg will be our search example for this chapter. He is famous for his invention of Cornflakes breakfast cereal with his brother and is known for his eccentric ideas and methodologies. He advocated for a vegetarian diet, extolled the virtues of celibacy, and supported eugenics programs and racial segregation.

The first step in the information gathering process is to gather background information about your topic and identify your search terms. One of the first search results for John Harvey Kellogg is for the online encyclopedia *Wikipedia*. The online encyclopedia entry provides important dates, places, historical information, names of persons involved, names of organizations, government agencies, societies, membership, associations, and affiliations. A researcher can use these to locate institutions and archives. A search string of "John Harvey Kellogg Seventh Adventist Church archive" returned multiple archives that contain correspondence between Kellogg and various church members.

Wikipedia entries often mention primary sources within the text, and these sources are usually hyperlinked to digitized materials. Otherwise, citations can be easily found using the Google search engine to see if the source has been digitized or is in the public domain. A digitized copy of Kellogg's publication *Plain Facts for Old and Young* can be located through Google in Project Gutenberg, the University of Virginia Library, and the Internet Archive.

Google Books provides a valuable resource that allows a researcher the ability to search the full text of any books and magazines that Google has stored within its database. When searching for primary sources, it is often beneficial to look at the bibliography of a secondary source. Once you have found a source, Google the source to see if it has been digitized or is in the public domain. Public domain books will often be located in Project Guttenberg and Google Books.

Library pathfinders are detailed web-based tools that clearly and concisely point users to helpful guides and are another useful source of information. Often created by library professionals, these research tools give you an overview of the resources from subject experts. They link to resources that are both digital and print within the library's collection as well as freely accessible resources found on the Internet. One of the benefits of a library guide is that it primarily assists in finding and discriminating among the resources it highlights. Library guides also help promote unique collections. They call attention to and highlight underutilized resources that the average person might not know existed. An example of a search string query to find such items would be "John Harvey Kellogg libguide." "Libguide" refers to one of the most popular web publishing platforms for libraries.

Sites often post images that do not link back to the original site. It is extremely easy to track down the original image using Google's Image search. From the Image search page, click on the small camera icon on the far right side of the search bar. To access it, drag and drop an image onto the search bar, upload an image from your computer, or right click the image and copy the URL into the search bar. This search technique is useful in a variety of ways. By searching the picture of a building, Google will ascertain the name and history of the building. Additionally, you might be able to find a higher-quality image, which may make it better for readability or presentations. However, Google is not always successful because their search algorithm will return results that contain visually similar images, color themes, and elements that are not related to the original image. Using a descriptor in the search will help return more relevant results.

Lastly, many researchers find broken or disabled links in the course of their research. A dead link does not mean that an item has vanished from the web. Often links are updated or moved, and the Wayback Machine, a digital archive of the World Wide Web and other information on the Internet created by the Internet Archive, will often provide a copy of the information. Additionally, the Wayback Machine may have additional archived resources that a user can obtain through the search interface.

SEARCH OPERATORS

Google uses search operators in much the same way as traditional databases. One tactic is to use Boolean Logic. This is a type of search that allows users to combine search terms using operators such as AND, OR, and NOT.

In a traditional database a query might be *"John Harvey Kellogg" AND "vegetarian diet."* This approach will limit search results to documents containing these two search terms. However, Google will automatically place an AND between the terms.

An OR operator between terms will return pages that have at least one of the terms. With this approach it is helpful to list synonyms separately. Common synonyms while searching for primary sources include diary, source, narrative, account, firsthand, first person, document, documentary history, oral history, interview, autobiography, charters, correspondence, manuscripts, pamphlets, personal narratives, and speeches letters. For example, *"first person" OR "document" OR "documentary history."*

To exclude terms using a Boolean NOT operator, add a minus sign (-) before the terms. This will narrow the search by requiring the first search term while excluding results that feature the second terms. This strategy is helpful when the second term is commonly associated with the previous term. Researchers should use caution as this search strategy may eliminate beneficial search results. For example, *"Higher education" -"community colleges."*

In a traditional database, truncation allows for some flexibility while searching. It allows the user to search for multiple forms of a word. By placing an asterisk at the end of a word, the database will look for variant endings. For example, diar* will return diary and diaries. However, Google will automatically search for variant endings of some words.

To search within a site or domain, use the search string *site:query* . This type of search is helpful if you are looking for more results within a particular website. For example, a search of " John Harvey Kellogg" on the *New York Times* website would look like *"John Harvey Kellogg" site:nytimes.com* . Finally, use quotes to search for an exact word or phrase. The advanced search page can assist you if you don't remember all of the operators and search strategies.

OTHER SEARCH TIPS USING GOOGLE

- ~manuscript (use ~ for synonyms)
- John +Harvey +Kellogg (use + to include the exact form of the word)
- define:archive (use *define:* for definitions)
- John * Kellogg (use the asterisk to replace a word or an initial)

GOOGLE SERVICES AND TOOLS

All of the following Google services and tools can be used along with their search engine to find primary sources and secondary resources.

- Google Scholar combines the user-friendly Google search engine with access to scholarly papers. Users have the ability to search across multiple disciplines in one search. However, users will have to evaluate the content for reliability and accuracy.
- Google Books searches the full text of books digitized by Google. It allows the user to find rare and out of print books, and as long as they are within the public domain, users can access the full text.
- Google Patents searches the full text of United States patents.
- Google Image Search searches for images on the web and provides users with the ability to search through the archives of *Life* magazine.

EVALUATING CONTENT

One of the disadvantages of using the Internet is that the sites harvested by Google during its search process are not regulated for quality or accuracy. Anyone can publish public opinions, hoaxes, and satire. Additionally, many sites publish facts without citing their original sources, which makes the information difficult to verify and evaluate.

Domain suffix can give you clues about the origin of a website. In general, it is best to look for websites that offer nonbiased and balanced informational resources.

- *.com* is a business or commercial site. These for-profit sites are helpful in the information gathering process, but a scholar would need to find the original source through a more reputable location, such as a traditional library database, archives, or an institution.
- *.edu* is for an educational institution. Schools from kindergarten through higher education utilize this domain suffix. Researchers should pay attention to see if the content is from a department or research center, or personal student and faculty pages.

- *.gov* is a government source and considered to be credible.
- *.org* is a nonprofit or advocacy website. Examples include the National Rifle Association (NRA), Greenpeace, or the Sierra Club. Some sites advocate a particular point of view, which can be either mainstream or radical.
- *.mil* is a domain suffix utilized by various branches of the Armed Forces of America and is considered to be a reliable source.
- *.net* was originally assigned to network providers. Scholars should verify information through a secondary source.

Coverage

- Google may return results that provide uneven coverage of a topic. The Internet Archive coverage favors domestic websites while providing uneven coverage of international sites.

Currency

- The issue of currency is not a problem for primary sources due to their historical nature. Additionally, a well-maintained website and current and functional links do not play a role in the validity and authenticity of information.

Authority

- Most archives and institutions will have a place on the website where they state their mission or vision statement. Most collections have been vetted for authenticity. Some questions to consider: Are they presenting information from a particular point of view? Which organization or archive is making this material available?

Google will often take you to major collections of primary sources. These collections are known for their authority. Additionally, many of these sites utilize a Google search bar to search their contents.

- American Memory presents collections from the Library of Congress.
- Avalon Project from Yale Law School documents law history and diplomacy.
- Documenting the American South is an archive from the University of North Carolina at Chapel Hill relating to the American South.
- EuroDos explores primary historical documents for European countries and regions.

- Gallica is the digital library of France.
- Internet Archive is a 501(c)(3) nonprofit organization whose goal is to index the web.
- Internet History Sourcebooks Project from Fordham University is a collection of public domain and copy-permitted historical texts.
- Making America focuses on American social history from the antebellum period through Reconstruction.
- National Archives and Records Administration (NARA) is the archive of the U.S. government.
- New York Public Library has a vast collection of digital images.
- Perseus is a collection of online resources from Tufts University for studying the ancient world.
- The Smithsonian Institution offers primary sources and other resources on a variety of topics.

ARCHIVES AND FINDING AIDS

The Google search engine is great for finding citations and finding aids; however, if these items or materials are not yet digitized or still within the copyright period, a scholar will have to visit the archive or submit a request. Some items are easily obtained through interlibrary loan while other rare and unique items must be accessed directly through an archive.

The cost of travel and institutional research costs may be cost prohibitive to some users. An archive will often charge for research assistance, photocopying, room use, day passes, and microfilm duplication, and have other processing fees. Some items are so rare and fragile that institutions delicately balance providing access with the interests of preservation. Lastly, the technical language and a closed stack environment of an archive might be a barrier to the casual researcher.

CONCLUSION

In terms of primary sources, the researcher is not looking for a broad overview but rather very specific and unique information. The Google search algorithm returns a large number of records and can return results that do not relate to the subject. Some entries are redundant and not easy to decipher. Additionally, Google offers limited search options that do not make full use of the subject-rich descriptors. This might prevent the user from working efficiently, instead making them scan and skim.

Currency of information is another limiting factor. Google does not offer a publisher list, journal list, time span, or disciplinary distribution. There is also a lack of bibliographic control leading to misspelled author names and a lack of standardization in titles. Some titles are presented with their full title while others are abbreviated.

The success of the search will depend on coverage, currency, and the search algorithm. Sometimes there will be no locatable primary sources for a given topic. The topic may be obscure, materials may not be digitized, or there may be no archival finding aid. Despite these limitations, the mixing of scholarly works, repositories, and other information sources make Google a quick and efficient discovery aid for primary sources. One of the benefits is that mass digitization makes source material readily available that would otherwise be costly, time-consuming, and difficult to access. Google should be used in conjunction with traditional library resources as an information recovery tool to maximize search results.

Part II

Collaborations

Chapter Seven

Hanging Out with Google

Amy James and Robbie Bolton

In 2011, Google gave up on yet another failed attempt at building a social network, replacing Google Buzz with Google+. Google has had a healthy list of failed forays into the social network market (McCracken 2014). Even though the reviews of Google+ have been mixed, at best, it has survived for three years now. Google+ claims a half billion active monthly users but most estimate the number to be closer to half of that (Miller 2014). With more than a billion active monthly users, Google+ is not much of a threat to Facebook's dominance of the social media market (Ha 2014). However, some have speculated that challenging the social network market is not Google's endgame (Miller 2014). The merits of Google+ can be debated, but the brightest star of all the features is its video conferencing tool, Google Hangouts.

Google Hangouts is a video chat and web conferencing tool anyone with a Google+ account (or Gmail account) can use. It provides many of the features of fee-based video conferencing tools without the cost. Hosting a Google Hangouts session is affordable for all libraries. It does not require downloading and installing software. It allows for a group video chat with multiple users simultaneously. Users can also be dialed in by phone if they don't have access to a computer or if they don't have a Google+ account (*Official Gmail Blog* 2013). The ubiquity of those with a Google account— Gmail claims 425 million active monthly users—ensures many patrons and librarians already have access to this tool without having to create a new account (D'Orazio 2014). The real value of Google Hangouts is the integration of Google Drive documents and the many useful apps available in the Hangout Toolbox.

In this chapter, we will briefly discuss the features of Google Hangouts and compare it with other video conferencing tools along with the advantages and disadvantages of using Google Hangouts. This chapter is not a "how-to" for Hangouts—there are plenty of resources on how to get started and make the most of your Hangouts experience (McKinney 2013). Instead we will provide examples and ideas for how libraries could use Google Hangouts for marketing and outreach, broadcasting live events, and library instruction.

ALTERNATIVES TO GOOGLE HANGOUTS

Why would a library choose to use Google Hangouts over other similar products? GoToMeeting, Skype, and JoinMe, are a few of the vendors that offer similar products to Google Hangouts. GoToMeeting is the product that most closely resembles all of the features and tools of Google Hangouts. Skype is best known for instant messaging and video chat. JoinMe is a free tool that offers easy screen sharing and remote desktop control.

One of the main reasons Google Hangouts is preferred over some of these competing products is cost. Google Hangouts is free. To have access to some of the same tools and features provided by Google Hangouts, GoToMeeting charges a subscription fee. To have access to similar features on Skype, you must be a premium subscriber. Although recently Skype has announced that they are allowing group video chats for nonpremium subscribers (Chacos 2014). Another advantage of Google Hangouts over these other products is the sheer ubiquity of Google. We have already established that there are about 425 million active monthly users. Skype is the only one of the mentioned competitors that could rival this, with Microsoft reporting Skype has approximately 300 million active monthly users (Perez 2013). While Skype has a wide reach, it offers fewer tools as part of its free products.

Some fee-based services might offer a slightly better overall video conferencing software, but Google has put together a full suite of tools (multiple simultaneous participants, broadcast capabilities, screen sharing, etc.) and a large user base already familiar with its products that cannot be rivaled by another company for a cost even the most budget-conscious library can afford.

GOOGLE HANGOUTS FOR LIBRARY MARKETING

Businesses small and large, musicians, astronauts, and even the president of the United States of America have all used Google Hangouts at one point or another to connect with their target audience. A target audience is the group of people for which a business or company has strategically decided to market their services. Connecting with the target audience is important for a number of reasons, but one of those reasons is to hone in on what your target audience is looking for from your company so that you can better market your products and resources to them. For example, academic libraries are almost exclusively marketing their services toward the needs of their students, whose tuition dollars go toward purchasing the library's collections and supporting their efforts. Public libraries market their services toward the taxpayers who live in the communities that they serve, and school libraries focus on marketing their services toward their students and teachers for educational purposes. Each library type can benefit from using Google Hangouts to market their collections, events, and resources. Google Hangouts provides a unique opportunity for libraries to reach out to their users and can be a great way to create or amp up marketing strategies.

In an article titled "We Should Hangout," author Russ Martin talks about how media brands and publishers are using Google Hangouts "as an editorial tool and way to reach a level of engagement that's difficult to achieve with traditional media" (Martin 2013). He also discusses how companies are using Hangouts to solve customer service–related issues and even invite select users or customers to participate in a Q&A Hangout session. Ultimately, the tool is being used to drive customer engagement (Martin 2013). Gretchen Howard, global head of sales and strategy for Google+, says Hangouts are a way to "augment the content they [companies and businesses] already have and go deeper with their customer needs" (Martin 2013). So, how does that translate to nonprofit organizations such as libraries?

Marketing, for libraries, is not just about selling a product. Outside of marketing physical materials and events, libraries are oftentimes promoting larger goals and objectives. For example, one of the overarching goals and missions of libraries in a general sense is to promote education and lifelong learning. *Teacher Librarian* and active *School Library Journal* member, speaker, and presenter Shannon Miller markets specific books, authors, and even special events in support of this goal. Miller set up Google Hangouts and Skype sessions to promote and support World Read Aloud Day and World Dot Day, "inspired by Peter Reynolds's book *The Dot* (Candlewick, 2003), about student creativity and affirmation" (Miller 2014). In doing so, she not only helped get the word out about specific books, authors, and worldwide events, but she also contributed to the promotion of reading and lifelong learning. Translating Shannon Miller's strategy into the world of academic libraries is not too tough a feat. Consider using Google Hangouts On Air to do a brief weekly promotion of collections or materials. If, for instance, there is a large project that many students are working on, hosting a live Hangouts On Air session to promote parts of the collection (print or electronic) that provide support for that assignment would be a useful way to help students and market materials.

Promoting library events is one of the other areas that Miller focused on in her Google Hangouts and Skype sessions. One way to continuously promote different events (either community events, library-specific events, or larger world-wide events) is to use Google Hangouts as an outlet to host public conferences or meetings where community members have the opportunity to share ideas and comments about past, present, or future library events. Academic libraries could use Hangouts in a similar way, by hosting public meetings and allowing students to share suggestions about library activities or events. Libraries could also use Hangouts On Air to host live webcasts of events as they are happening. The events would then be available on YouTube for those who weren't able to attend. Spring Arbor University Library has done this for university events. The university hosts Community of Learners events one Friday a month during the academic year. This event consists of faculty giving talks on academic topics in their areas of concentration. The series focuses on lifelong learning and the importance of education and research. The library helps provide access and promotes these events to on-campus and off-site students by hosting them as a Hangouts On Air session. This allows students to attend virtually if they cannot be there in person.

Besides marketing collections and events, it is also important for libraries to market the services that they provide. Of the many services that libraries supply, reference and library instruction are some of the most crucial. This is the area where the use of Google Hangouts in libraries really shines. Imagine marketing your library's reference and research instruction services by actually hosting live On Air instruction sessions. The sessions wouldn't necessarily have to be geared toward one particular class, but they could focus on general information literacy skills, or they could be focused on higher-level research skills. Individuals in classes that need or want extra help could set up sessions with small groups of students working on projects. What better way to promote your research instruction services than actually hosting them live?

HANGOUTS ON AIR

Hangouts On Air is an easy way to broadcast or live-stream a Google Hangouts session. As the Google Hangouts session is beginning, the simple click of a button will stream your session to your patrons and followers. Broadcasting your Google Hangouts session allows anyone online to watch your session in real time (or on a slight delay). Send a link to the live session to your patrons or embed it on your library website to allow a much wider audience to engage with your library. Hangouts On Air automatically creates a recording of your Hangouts session and posts it to your YouTube channel for easy viewing.

Google provides an optional Hangout Toolbox app that can be used with your Hangouts On Air sessions. Some of the features included in the toolbox allow you to place banners at the bottom of the screen to display the presenter's name and title, or any text for that matter. Another essential toolbox feature is the comment tracker, which allows your audience to actively engage with your presentation by posting comments and questions.

Some of the ways that libraries are using Hangouts On Air include streaming live events and lectures, hosting professional development webinars, and even making quick library instruction videos and tutorials. As previously mentioned, Hangouts On Air have been used to host Community of Learners lecture series events at Spring Arbor University. Most types of libraries host events at the library, whether lectures, poetry readings, book clubs, and so forth. One elementary school library is using Google Hangouts as a way for students to interact with authors (Ruth Borchardt Elementary Library 2014). Hangouts On Air is also an excellent tool for professional development opportunities and collaborating with other libraries. The American Libraries Live is doing great work in broadcasting topical conversations with librarians using Hangouts On Air (Ala TechSource 2014). Broadcasting a Hangouts On Air session is also a very quick and informal way to record a video to share with patrons or to create a screencast.

GOOGLE HANGOUTS FOR INSTRUCTION

Using Google Hangouts to host live instruction sessions can be a great way to connect with your students. In an article called "'Hangout' with Your Students Using Google," authors Caroline Haebig and David Lawrence talk about the benefits of using Google Hangouts to connect with students. They point out that using "Google+ Hangouts video conferencing allows users to simultaneously collaborate on a Google Doc, view a YouTube video, and use screen share and other apps . . . a powerful way to capture collaborative efforts and conversations!" (Haebig and Lawrence 2013–2014). The Hangouts On Air function allows for an unlimited number of people to view the session, and there are also capabilities to include comments and questions. These features support library instruction sessions because they allow for real-time remote communication at no cost.

In a recent study called *Creating a Virtual Academic Community for STEM Students*, Google Hangouts were used for remote tutoring with students in a science, technology, engineering, or mathematics field of study, specifically concentrating on students who were deaf and/or hard of hearing. The tutoring sessions were conducted synchronously and remotely. Results from the study show that "respondents felt that remote tutoring compared favorably to in-person tutoring" (Elliot et al. 2013). Results also showed that tutors and students felt it was easier to schedule times for tutoring because they could conduct the sessions remotely. The only issue that students found with conducting the tutoring sessions using Google Hangouts had to do with technical issues, unstable video quality, and problems understanding the frequent changes in the Google+ app. However, the overall results proved favorable. Libraries can use this study as an example for conducting similar sessions, but for library and research instruction purposes rather than tutoring purposes.

If instruction sessions are not hosted using Hangouts On Air (making them publicly visible with an unlimited number of audience members), they could be hosted regularly with up to ten students at once. This would be a great way to reach out and provide instruction to upper-level and graduate classes (which tend to have a smaller number of enrolled students), or to work with small groups of students or offer one-on-one research instruction.

GOOGLE HANGOUTS FOR PERSONAL LIBRARIAN SESSIONS

One-on-one instruction using Google Hangouts can be useful for libraries that currently have, or are interested in creating, personal librarian programs. In a personal librarian program, subject librarians are promoted to students as their "personal librarian," which encourages students to see the librarian as approachable and as someone who they reach out to on campus (Henry, Vardeman, and Syma 2012). Several libraries have taken on personal librarian campaigns to market the subject librarians' research services to students. One library that has had success with the implementation of a personal librarian program is Texas Tech University. Texas Tech University began filming brief two-minute videos of each personal librarian introducing him- or herself. The librarians discussed their hobbies and more lighthearted topics in order to make them seem "more personable, authentic, and 'real' to students" (Henry, Vardeman, and Syma 2012). Their goal was to actually market the librarian as a "person."

Google Hangouts would be a beneficial tool to use for personal librarian programs. Rather than creating introduction videos, librarians could do a Hangouts On Air session for each subject area's personal librarian. Students in each particular field of study could attend their personal librarian's introduction session via Google Hangouts On Air. Using the question and answer add-on, participants in the session would have the opportunity not only to meet their personal librarian, but also to ask questions about research generally or in their major or for specific projects that they may be working on and get answers to any other questions that they might have.

Using Google Hangouts, the personal librarian program could be extended beyond traditional students. It would have the potential to reach online and distance students. Librarians would have the opportunity to conduct Google Hangouts introductions and then later meet with groups or individual students conducting research in their field of study. For example, if a college or university had a cohort of students in a business program that were all completing the program as distance students, librarians could connect with the advisors and instructors of those programs and set up virtual Google Hangouts with those students. Initially, the introductions and larger group meetings could be held via Hangouts On Air so that multiple students could attend. Then, when students want to meet with their personal librarian to get additional help, librarians that are on the main campus would be able to schedule a Hangout with students and have individual meetings with them regarding research in their major.

CAVEATS AND CONCLUSIONS

Google Hangouts offers many unique features that set it apart from other video conferencing tools. Not only is it a free service, but it also provides the functionality of many of the paid conferencing software. Other companies have caught onto the need for group video chats. Skype now offers group video chat services to nonpremium subscribers. However, users will need to abide by the restrictions on call length and total video call hours set by Skype (Skype 2014). Google Hangouts does not set these kinds of restrictions on users. Despite the benefits to Google Hangouts, there are still some things to watch out for with the service. One concern with Google Hangouts is that users must have a Google+ or Gmail account in order to access the Hangouts service. If users don't want to create an account with the social network or e-mail provider, they will be unable to conduct a Hangouts session. However, users are still able to view a Hangouts On Air session even if they don't have a Google+ account because of its visibility on YouTube. Another concern with Google Hangouts is that you can only have a group hangout with up to ten people at a time (Google 2014). Again, using a Hangouts On Air session with the question and answer add-on can be a helpful way to get around this. Even though there are some downfalls with using Google Hangouts, the positive attributes certainly outweigh the negatives. With the ability to use the service to market your library, host Hangouts On Air for events, teach library instruction sessions, and even create a personal librarian program for distance and online students, Google Hangouts is definitely a tool worth checking out.

REFERENCES

Ala TechSource. 2014. "AL Live: The Copyright Conundrum." http://www.youtube.com/watch?v=qR_gNQB2RbE.

Bell, Steven J. 2011. "A Conference Wherever You Are: The Virtual Conference Experience Improves—but Don't Write Off the Real Thing Just Yet." *Library Journal* 136 (16): 28–31.

Byrne, Richard. 2012. "Making the Most of Video in the Classroom." *School Library Journal* 58 (8): 15.

Chacos, Brad. 2014. "Watch Out Google Hangouts: Skype Rolls Out Free Group Video Calling." *PC World*. http://www.pcworld.com/article/2148666/watch-out-google-hangouts-skype-rolls-out-free-group-video-calling.html.

Dockterman, Eliana. 2013. "Never Visit a College Campus with Your Parents Again Thanks to Google." http://nation.time.com/2013/10/24/never-visit-a-college-campus-with-your-parents-again-thanks-to-google/.

D'Orazio, Dante. 2014. "Gmail Now Has 425 Million Active Users." *Verge*. http://www.theverge.com/2012/6/28/3123643/gmail-425-million-total-users.

Elliot, Lisa B., Benjamin Rubin, James J. DeCaro, E. William Clymer, Kathy Earp, and Michele D. Fish. 2013. "Creating a Virtual Academic Community for STEM Students." *Journal of Applied Research in Higher Education* 5 (2): 173–88.

Google. 2014. "Google Hangouts Help." Accessed April 25, 2014. https://support.google.com/hangouts/answer/3111943?hl=en.

Ha, Anthony. 2014. "Facebook Says It Has 1B Monthly Active Users on Mobile, 200M Actives on Instagram." *TechCrunch*. http://techcrunch.com/2014/03/25/million-vs-billion-joke-goes-here/.

Haebig, Caroline, and David Lawrence. 2013–2014. "'Hangout' with Your Students Using Google." *Learning & Leading with Technology*, December/January: 26–28.

Henry, Cynthia L., Kimberly K. Vardeman, and Carrye K. Syma. 2012. "Reaching Out: Connecting Students to Their Personal Librarian." *Reference Services Review* 40 (3): 396–407.

"Kerala CM to Connect with over Two Million Students on Google Hangout." 2013. *PC Quest*.

Martin, Russ. 2013. "We Should Hangout." *Marketing* 56.

McCracken, Harry. 2014. "A Brief History of Google's Social Networking Flops." *Time*. http://techland.time.com/2011/07/11/a-brief-history-of-googles-social-networking-flops/.

McKinney, Debby. 2013. *Google Hangout Marketing the How and Why*. Kindle edition.

Miller, Claire Cain. 2014. "The Plus in Google Plus? It's Mostly for Google." *New York Times*, February 14. http://www.nytimes.com/2014/02/15/technology/the-plus-in-google-plus-its-mostly-for-google.html.

Miller, Shannon McClintock. 2014. "Innovators: Virtually Connected." *Library Journal* 139 (5): 50.

"NASA to Host Google+ Hangout on Hurricane Research Flights." 2013. *PR Newswire*.

"Official Gmail Blog: Making Calls from Hangouts—in Gmail and across the Web." 2014. *Official Gmail Blog*. Accessed April 28, 2014. http://gmailblog.blogspot.com/2013/07/making-calls-from-hangouts-in-gmail-and.html.

Pham, Alex. 2012. "Hanging Out with the Band." *Billboard* 124 (40): 26–27.

Perez, Juan Carlos. 2013. "Lync-Skype Integration Live Now Worldwide." *Network World* 30 (10): 18.

Ruth Borchardt Elementary Library. 2014. "5th Grade Ellen Potter Author Google Hangout." Accessed April 28, 2014. http://borchardtlibrary.edublogs.org/2014/02/03/5th-grade-ellen-potter-author-google-hangout/.

Skype. 2014. "Group Video Calling Fair Usage." Accessed April 28, 2014. http://www.skype.com/en/legal/gvc-fair-usage/.

"Virtual Reference and In-Depth Assistance Using Shared Workspaces." *Online Searcher* 37 (1): 22–26.

Weiss, Todd R. 2014. "Google+ Hangouts to Host Virtual Road Trip with President Obama." *Eweek* 3.

Chapter Eight

Harnessing the Power of Google Docs for Writing Collaboration

Jaena Alabi and William H. Weare Jr.

INTRODUCTION

Librarianship is a remarkably collaborative field. In addition to working with colleagues at our own institutions, sometimes we collaborate with colleagues across the state, country, or globe. In order for these cooperative efforts to work, we need a reliable, versatile, and accessible tool that facilitates collaboration. Google Drive provides one such tool in its Docs application.

Drive is Google's suite of office tools (like Microsoft Office) combined with sharing capabilities (akin to Dropbox). Drive allows users to create, store, share, synchronize, and edit files. Each application in Drive can be used to create different kinds of products: documents, spreadsheets, presentations, online forms, and even drawings. We will focus on Drive's word-processing application, Google Docs. It is much like any other word-processing program, except that it lives in the cloud rather than on the desktop. This allows you to access your documents wherever you have an Internet connection, which makes it easy to collaborate in real time from anywhere in the world.

As academic librarians, both of us have used Google Docs for a variety of purposes: communicating ideas with others, soliciting feedback on a document, sharing articles found during research, distributing documents to committee members, and circulating meeting minutes. While these storage and sharing functions do make it easier to disseminate information, they are activities that fall short of actually writing—getting words on a page.

When we first started collaborating, we did not use Google Docs to its full potential. After some experimentation, however, we have figured out how to use Docs more effectively, and we now use it to write together in real time even though we live in different states. This new approach has made us more productive. In this chapter, we will describe how we write together using Docs and share strategies for using this tool to get things done.

HOW IT WORKS

To access Google Docs, go to www.drive.google.com (or docs.google.com , which will redirect you to the Google Drive sign-in page). If you do not already have a Google or Gmail account, you will need to create one. After you have created an account and logged in, you can create a new document or upload an existing file.

In general, Google Docs operates much as Microsoft Word does. One obvious difference, though, is that many of the formatting options in Docs are tucked away in the toolbar and drop-down menus instead of being displayed in a large ribbon as they are in Word. Also, unlike Word, which prompts you to save a file when exiting, Docs saves automatically as changes are made to the document. The simplicity of design found in Docs helps the user focus on actually writing rather than wrestling with layout issues.

When we started using Google Docs to collaborate, there were a number of functions that we had grown accustomed to in Word that were not available. We have occasionally been surprised, though, when some functionality that had been missing is now suddenly available. For instance, while writing this chapter, we lamented the lack of a thesaurus tool in Docs; however, there is now a thesaurus add-on available for the program. Coping with a product that is in perpetual beta—where new developments and improvements are continually being added—is part of using Docs.

Google Docs is more useful than Word for collaborative work because it makes it easy for you to share. Instead of crowding others' inboxes with attachments, simply click on the "share" option and enter in your collaborators' e-mail addresses. Next to each e-mail address, you will find a drop-down box that you can use to determine what rights that person is allowed: if they can simply view the document, if they are allowed to make comments on the document, or if they have full editing privileges.

When your collaborators follow the link from the e-mail they receive from Drive, they will be able to log in to your document and view, comment, or edit. When you and your collaborators are logged in to the document at the same time, each of you will see one another's initial inscribed in a small, colored block at the top of the screen. Each person's cursor will be visible to the other collaborators in a color that corresponds with their colored block. If everyone has editing rights, you will also be able to see your fellow writers highlighting, adding, or deleting text, all in real time.

Google Docs also has a chat function. By default, the chat box appears in a corner of the same window as the document, but it can be "popped out" or opened in a new browser window. This chat function allows you to talk about the writing process with your collaborator in a separate space, leaving the document itself for the actual writing (though that is not always what we do—more on that later).

WHY IT CAN WORK FOR YOU

The design and features of Google Docs makes it easier to collaborate in real time. Working simultaneously enables each writer to build on the other's ideas. When you have a collaborator, you are both writing, but you are also both serving as readers for the other and can thus identify points that need clarification. It also helps ensure that collaborators are working toward the same end, rather than discovering at a later time that each has gone in a different direction. Additionally, working with a partner can foster a sense of accountability; we have both found that it is harder to put aside or give up on a project when that would mean letting someone else down.

When collaborating, there are several things you and your partner(s) should pay attention to, including work and writing preferences. Perhaps one writer works in a fast-and-furious fashion, while the other takes a slow, methodical approach to writing. Or one person may prefer to get a rough draft down on paper while another might prefer to edit the writing of others. Some writers prefer to work every day for relatively short periods of time; others prefer to dedicate larger blocks of time less frequently to a specific project. Having an understanding of the way you and your collaborator(s) prefer to work may help you address the tensions that can arise from conflicting work styles.

One of the challenges in any collaboration is finding time to work together. As most librarians have busy schedules, it can be difficult for two (or more) of us to schedule a time to meet and work together online. Writing collaboratively requires both energy and focus, which, for many people, can vary according to the time of day. It might also be difficult to find a time when both are focused and productive—especially if one writer is a morning person and the other is not. Compromise will likely be necessary. In addition to selecting a time of day that works for both, we recommend meeting at roughly the same time each day to help establish a writing routine.

It is important to be sensitive about the work of others and to offer suggestions with tact. Some writers are attached to their writing, interpreting any critique of their text as a personal attack. Others, in an attempt to avoid offending a coauthor, may hold back on suggesting changes. Withholding such constructive critical feedback lessens the benefits of collaboration.

When collaborating, it is crucial that the parties involved be detached enough that they can accept criticism while also being sufficiently invested to suggest improvements. This balancing act is more difficult when relying on chat or e-mail to communicate because such messages, in the absence of tone of voice and body language, are easily misinterpreted.

HOW IT WORKS FOR US

In many situations where people are working together in what they consider to be a collaboration, the labor is often divided—"I'll do this part. You do that part. Then, we'll put the pieces together." For us, however, collaboration means truly writing together. From outlining to drafting, from revising to editing, we usually work on the same task at the same time. This allows us to play off one another's ideas and suggest improvements to the other along the way. We have found this approach to be successful for us because our strengths and abilities complement one another.

Our work styles and preferences vary from day to day and project to project. For some projects, we like to work quickly, sketching out a rough draft and revising extensively later. Other times we focus on getting it right from the start; this can be slow, but requires less revision. On days when one of us is feeling sluggish, the other will often take the lead. Sometimes one of us will focus on cleaning up the document—removing extraneous comments, correcting the other person's punctuation or spelling, or suggesting alternate words. Though not particularly onerous, these activities help to move the project forward.

While neither of us is especially energized in the afternoon, there is something about working collaboratively—being online and having to respond to one another's writing and comments in real time—that keeps us focused and engaged in the process. Meeting in the afternoon enables each of us to focus on other commitments and individual projects earlier in the day. It also helps both of us overcome what would otherwise be a late-afternoon lull in productivity; we accomplish more working together than we would alone.

On earlier projects, we worked one or two days per week for an hour and a half or longer, but we now work every day—or nearly so—usually for an hour. Initially we were concerned about the brevity of our meetings. We were not sure that we would recall where we had left off or that we could be productive given the short time period. We have managed to make it work, though, by leaving notes to ourselves at the end of each session. These notes remind us where we left off and what task needs to be completed next.

We have found that working with another person provides a level of accountability that has positively contributed to our productivity. The times that we reserve on our calendars to work with one another online are like formal appointments. Having found that it is easy to put aside individual projects when something seemingly more important comes up, we are less inclined to do this when we make an appointment to work with each other. In order to avoid disappointing our partner, each of us shows up—virtually and mentally—at the designated time.

We have also learned that working with another person requires a combination of tact, willingness to compromise, and patience. Tact is essential when recommending changes to another person's writing. For example, when we share thoughts on sentence order, word choice, and other issues, we do not say, "Ick! That's terrible." Instead, we use a more positive phrase, such as, "I have an idea"—and offer an alternative. We also use direct and specific comments: "That phrase—the one I've just highlighted—I think it might be too informal." The willingness to compromise is particularly important when it comes to accepting suggestions. One of us might question a change, which sometimes leads to a prolonged discussion. Other times, one of us simply yields, especially if neither has a strong preference. Collaboration also requires patience. On any given day, either one of us (sometimes both) might be having a difficult time concentrating. We both understand that there are other projects that require our time, and we have colleagues who may need our attention as well. It helps to keep in mind that the goal is completion (and subsequent submission) rather than perfection. Finally, in addition to tact, a willingness to compromise, and patience, a sense of humor is beneficial.

HOW TO MAKE IT WORK FOR YOU

Having worked together on several projects over the past three years, we have developed a number of work habits that have made the collaborative process easier. We start by talking through our ideas on a topic and agree upon a direction. Then we create a shared document, craft a thesis together, and draft an outline. Early on, we do a lot of our chatting directly in the document, rather than in the chat box. This allows us to save our conversation and keep track of ideas as they develop. We often begin brainstorming and capturing thoughts, typing simultaneously, but not paying attention to what the other is doing. Both of us subscribe to the "write from the start" school of thought, so we put words on the page from day one.

Once we have some ideas down, we will look at what the other has typed, and might expand upon what is there, note additional ideas, or begin grouping similar thoughts. For some of our projects, we have taken turns creating sacrificial drafts of paragraphs or sections. Sacrificial drafts allow us to get started, knowing that what we have written will be torn apart and extensively revised. This approach enables us to complete a draft more quickly. In these early stages, we try to avoid getting caught up in editing or revising, instead focusing on capturing all of our thoughts for the project.

We often created and maintained several documents for one project when we first started writing together. We had one with a thesis and outline, another with the proposal or call for chapters, and yet a third with notes and source material. Now, however, we work in only one document. We initially post the proposal or call for chapters at the top of the document, followed by the thesis and outline. Notes and quotations from sources also go directly into this document (often in a different color). As our document begins to grow, we use the bookmarking and linking features available in Google Docs. We create preliminary headings for each section, bookmark them, and then link the bookmarks from our outline at the top of the document to the corresponding section. This makes it easier to move to specific sections of the draft and allows us to place ideas in their appropriate locations. Occasionally, we develop an idea into a sentence or paragraph but do not know where it belongs. We now keep a parking lot at the end of the document for such items.

In addition to creating an outline and provisional section headings, we use several other basic formatting devices—pagination, bold text, numbering, bullets, and brackets. Using such tools may seem simplistic, but they provide navigational markers and organization that are essential when writing collaboratively. Different font colors allow us to keep track of where we are in the document and show us what work has been done. For example, our first draft may be in purple, revised text may be in blue, and notes may be in black. Red, which is used sparingly, serves as a reminder that a particular item needs our attention.

Writing simultaneously in the same document can be challenging. For example, one person may move or delete a section that the other person is about to edit. To address this frustration, we have developed a routine for making revisions. We generally copy the draft text and paste it just below the original. We then revise the copied text, leaving the original intact. Sometimes we go through this process several times before arriving at a final revision of a section. Only after we have agreed on a final version do we delete the earlier draft. We retain the first draft in case we conclude that our revisions are not improvements over the original. This also allows us to go back and recover a particularly good word or phrase that may have been deleted in the revision process.

Our process of revision depends on the approach we have taken with that particular project. For some projects, we draft a portion on one day, and then on the next day we revise that section before moving on to the next piece. For other projects—like this book chapter—we focus on completing a full draft first, knowing that we will later complete a substantial revision. Alternatively, one of us might independently revise a paragraph and paste the suggested revision into the document, then we will review it together at our next meeting. No matter the approach, we regularly engage in discussions about such issues as word choice, grammar, sentence construction, and the order of paragraphs.

This process leads to a great deal of duplication in our working document—situations in which we have created multiple versions of paragraphs that address the same idea. Toward the end of each work session, we usually spend a few minutes cleaning up the document. We take turns: one person will go through the text and strikethrough what can be deleted, and the other person follows and either deletes the text or indicates that it should not be deleted yet—and why.

In communicating with each other, we rely heavily, though not solely, on the chat feature of Google Docs. Sometimes in the course of our work we realize we need to stop and have a conversation by phone, especially during the planning stage or when making extensive revisions. Talking with each other is also easier than typing when we need to rethink the direction of the article or work through a contentious point.

In addition to learning how to work with each other, we have had to learn how to cope with the occasional technology glitch. For instance, the chat box may indicate that one of us has left the chat. The person who has "left chat" will not know this though, and will continue sending chat messages. The other person will not respond because he or she is not receiving these messages. It usually takes a couple of minutes to realize what has happened. Closing and reopening the chat function does not fix the problem, but reloading the page usually does restore chat. We have also had issues with some functions when using the Firefox browser, such as copying and pasting. Now that we both use Chrome (Google's Internet browser) when working in Docs, we no longer encounter this problem.

CONCLUSION

We recognize that our approach to collaboration may not work for everyone. Nor is our approach particularly efficient—it seems to take more time to write with each other than it does alone. We find collaboration via Google Docs to be beneficial because it allows us to build upon each other's ideas, challenge each other, and support each other throughout the writing process. This kind of collaboration leads to a better final product. Together, we have completed a peer-reviewed article, a conference paper, two national presentations, and this chapter. Google Docs has been instrumental in making it possible for us to complete these collaborative projects.

Chapter Nine

Taking Interlibrary Loan Operations from Good to Great Using Google Collaboration Tools

Sarah Troy

The University of California, Santa Cruz (UCSC), began using Google e-mail as the campus e-mail system in Fall Quarter 2011. Over time, the campus also adopted Google applications, including Chat, Calendar, Drive, Groups, and Sites. Each of these applications brought improvements to existing practices, including increased efficiency and better knowledge sharing.

User Services & Resource Sharing (US&RS), known in most libraries as access services, manages most of the face-to-face public service in our two library buildings. US&RS includes circulation, collection maintenance (stacks), information services (reference desk), interlibrary loan, media center, and course reserves. US&RS accounts for nearly a third of library staff. The department oversees four physical public service points. On any given day, a US&RS staff member could potentially work from five different computer workstations. Because the US&RS existence is somewhat nomadic, having work live in the cloud makes the workday seamless and efficient.

Until recently, interlibrary loan (ILL) was unique within the US&RS department. All ILL assistants were classified at the same level, had identical job descriptions, and shared the same work. Staff members were fully cross-trained in lending and borrowing responsibilities. ILL documented all processes via workflow maps. Ongoing tasks rotated weekly or monthly. One of the greatest sources of pride for the department was that the work was not dependent upon one person. Any patron who called for assistance could get the same level of help, no matter who picked up the phone. This level of cross-training and communication allowed us to keep ILL running smoothly,

even when staffing briefly dropped from four to two full-time library assistants.

For a group of people to function efficiently, information sharing is key.
In ILL, information sharing took place in multiple venues. In addition to
weekly meetings, ILL had pick-up meetings whenever necessary, documented decisions and the reasons behind them, and communicated about
requests by placing notes in our ILL management software, Virtual Document eXchange (VDX). ILL had a culture of healthy communication. When
the campus adopted Google applications, it was clear that Google tools supported the spirit of collaboration that ILL engendered.

GMAIL

When Google e-mail (Gmail) was implemented at UCSC, all staff e-mail
accounts were automatically transferred from the campus e-mail system.
Departments were also able to create applications-only e-mail accounts to be
used as shared departmental e-mail accounts. These accounts can be set up so
that multiple employees have access to them. In ILL, library assistants and
student employees need to view and respond to e-mails from local patrons
and from other libraries.

Google allows for a lot of flexibility in e-mail settings. ILL appreciates
several of Gmail's features. Once an e-mail response has been sent, it is
automatically archived, and we can search for it and find it again if we need
to remember particular details. If patrons respond to archived e-mails, the
entire e-mail string returns to the inbox. Easy searching of e-mails allows
anyone in ILL to respond to inquiries and have all the relevant information
handy. Renewal requests from patrons are submitted online and automatically routed to the department's e-mail address. Because of the way Gmail
connects individual e-mails as conversations, many different patron renewal
requests arrive in one long e-mail string. Once we place all the renewal
requests, we delete the entire string. Any subsequent renewal requests that
come in will initiate an entirely new conversation string. The head of ILL set
up several filters that automatically route certain types of messages to specific folders. This allows ILL staff to skip the step of sorting messages in the
inbox. For example, all renewal requests are automatically sent to the renewal folder. All overdue notices are sent to an overdue folder.

Another convenience of Gmail is the canned response lab. Canned responses (a Google lab that has to be enabled) are e-mail templates. They are
easier to use than saving e-mail text in a shared Drive document because they
are right inside the new message field. While most of the notices we send to
our patrons are generated in VDX, occasionally there are times when we

have to initiate a message outside of that system. Canned responses ensure consistently phrased and polite messages for our patrons.

One difficulty with shared e-mail accounts is that our campus security recommendations prevent the e-mail account owner from sharing passwords. Anyone needing access to the e-mail account can be given delegate access. The list of delegates must be maintained so that when people leave their positions, it is kept up to date. Once delegate access was set up for the ILL e-mail account, staff discovered that every e-mail they sent listed their individual work e-mail address in the "from" address field. In ILL, we like to keep most e-mails anonymous to encourage patrons to work with anyone, rather than getting attached to a single library assistant. Additionally, we did not want our student employees to have their e-mail addresses sent to patrons. Using delegate accounts for e-mail is not our ideal scenario, but having the departmental e-mail account and all its conveniences outweighs other negative side effects.

CHAT

Google Chat is incredibly useful in day-to-day communications among ILL staff. There is an immediacy to instant messaging that makes it more efficient than a telephone conversation. For example, each week, the head of ILL and I have morning desk shifts at our two libraries. We often use this time to have brief, sporadic chats related to our ILL work. I may ask her something along the lines of "Did you catch that ILL number 12345 was finally returned? Don't bill the patron for it." And then a few minutes later "Can you do a shelf-check for me? Call number is QD123. . . ." And a few moments later "A patrons says he returned his ILL book at Science, can you check the return shelves for me?" These are conversations that are so quick and frequent that it would be inconvenient to talk over the phone. We are able to resolve issues and pass information along quickly via Chat.

As a department manager, I spend a lot of time in meetings. I may be away from my phone, but people still need to get in touch with me. Chat statuses allow me to show that I am busy, but staff members know that I am there and can be reached in a pinch. For times when I am truly not available, I go invisible on Chat. Group chats have also proved useful at times. All ILL staff used to attend quarterly system-wide conference calls. For the duration of those calls, the department opened a group chat window. We could ask questions of one another, get consensus for providing feedback on a topic at hand, and check for understanding when we heard things we could not interpret.

Interlibrary loan assistants do not have individual phone lines or individual offices. It is difficult to have private conversations. Chat offers an opportunity for us to ask one another questions or provide sensitive information discreetly. The discretion afforded by Chat also has safety benefits for public service desk workers. It allows us to contact one another without patrons at the desk overhearing our conversation. If I observe patrons behaving in ways that cause me concern, I can chat to my colleagues in their offices to ask for support at the desk or to request that they call for police assistance.

CALENDAR

The University of California system has ten campuses, plus the California Digital Library and two off-site storage facilities. When we have system-wide committees, scheduling meetings can be a chore. The beauty of a shared calendar is that you can more easily schedule a large group of people. While the UC system is not on a shared calendaring system (hooray for Doodle polls), Calendar is often smart enough to recognize time commitments in e-mails from UC colleagues, allowing the easy addition of items to the calendar. While we do not use Google Calendar UC system-wide, we do use it on the UCSC campus. Since there are many campus-wide committees and various meetings, it is a useful tool.

Library assistants use Google Calendar to schedule day-to-day commitments. Many library assistants in US&RS work nontraditional schedules. It is difficult to keep track of who is working at any given time. Calendar allows us to see easily who is in or out on a given day. Color coding options (e.g., unit meetings green, lunch breaks purple) make it easy to glance at your schedule and know exactly what you have for the day.

US&RS uses Google Calendar to schedule two circulation desks, an information services desk, and our Roving Student program. The beauty of using Calendar is that we can share all of our desk calendars with library staff, and we can embed them in public web pages. It also allows us to schedule repeated commitments, such as weekly meetings, as well as shared tasks. For example, every other week I am responsible for billing patrons for items they have not returned. I schedule a repeated reminder to do that task every other week. Calendar also has task lists to keep track of to-do items. You can add due dates to items and check them off your list when they are completed. You can create multiple task lists, so that you could potentially have a discrete list for each of your current projects. You can also e-mail your task list to yourself. Items with due dates appear both in the calendar and on the task list itself.

In addition to working on multiple public service desks, ILL staff also schedule shifts on a behind-the-scenes desk. Behind-the-scenes work involves answering the ILL public phone line and directing the work of student employees. The ILL desk is staffed by library assistants and by high-level student employees. Scheduling this desk can be a complicated mess, given the intricacies of student work schedules. Using Google Calendar, however, makes it easy to see when people are available.

ILL staff created a calendar titled "ILL Desk Availability." Each library assistant and eligible student employee enters their available work hours into the calendar. Meeting times and other commitments that preclude working desk shifts are excluded. Scheduling the ILL desk becomes much easier as everyone's availability stands out. We also use Calendar to manage student employee schedules, which change along with the academic quarters. Once the hours are entered, it is easy to see when additional student coverage is needed.

DRIVE

Long before UCSC adopted Google applications, coworkers and I used Google Docs for shared work. A colleague of mine and I used Google Docs to write an article together. The shared editing feature allowed us to draft portions individually and then edit one another's work and turn it into a seamless piece. The revision history allowed us to resurrect portions of text that we thought we did not want but ended up preferring. Google Drive, the successor to Google Docs, launched in April 2012. Google Drive serves as a hard drive in the Google cloud.

With Drive, work can be done from any workstation. If you are having a slow hour at the public service desk, you can work on a shared document. If you have to work from home for a day, you have access to all your files. US&RS worked on a department-wide project to identify out-of-print and rare media materials. Night staff led the efforts, but these staff members spend all their hours staffing the public services desks, and they do not have individual workstations. With Google Drive, they were able to update spreadsheets whenever the patron desk traffic would allow and from any workstation.

Google Drive has allowed ILL to create a shared brain. We store everything in Drive that anyone could potentially have a need to access. We recently had to edit all the automated notices that VDX sends our patrons. Rather than wordsmith over e-mail or in a meeting, we created Docs for each type of notice sent. ILL staff then edited the documents, leaving comments and highlighting changes.

In addition to larger projects, we use Google Drive for day-to-day operations. When we switched to FedEx for shipping, we dumped every bit of information about our FedEx account into a shared document. It is where we find our account numbers, primary contacts, login credentials, and instructions for ordering supplies. We have a similar document for our consortial courier service. We also store our list of reciprocal libraries in Drive. When we verify the picklist from any workstation, we can log in to confirm whether a library has a reciprocal relationship with us. When we bill our patrons for items they opt not to return, we access a shared document that tells us what various libraries will charge us. Google Drive helps ensure that ILL is not beholden to one person who has vital, unshared information.

ILL uses shared documents for meeting agendas and meeting minutes. Anyone can add items to the agenda, and minutes are added directly to the document, with the most recent at the top. Keeping all the minutes in one document makes it easy to search for text within the document and to find the minutes when you are in charge of taking them. One of our favorite uses of a shared document is our billing list. Each week we run a list of requests that are becoming very overdue and are in danger of being billed. We put information about all those requests in a spreadsheet, and we notify our patrons that their items are incredibly overdue. The following week, we use the document to bill patrons if those items have not been returned. Having all the ILL numbers in a spreadsheet allows us to easily copy and paste into our ILL software to see if the items have been returned. As we are checking to see what has been returned, we can also check the requests that coworkers are waiting for. This is all work that was previously done manually, with sheets of paper, which meant that if you were away from the ILL office, and away from the paperwork, you could not check on items that were soon to be billed. By putting the document in the cloud, we enabled ourselves to do the work from any location.

GROUPS

In addition to hosting e-mail reflectors for library groups, the library also hosts Google Groups as a favor for outside committees. UCSC library staff members serve on a number of consortial committees, and as a convenience for libraries with fewer resources, we provide a home for listservs. Once we began using Google applications, we migrated our e-mail reflectors. Creating and managing Google groups is not complicated. They allow departments to have more freedom in creating and managing reflectors, instead of relying on the campus IT department for support.

ILL has two groups: one for staff, and one for students. We use them to keep everyone in the loop about work issues, remind students to submit their timesheets, alert staff to software updates, and so on. Messages can be sent from an e-mail, or they can be initiated by logging in to an online forum. Messages are archived and easy to access. Not all members need or want to see every message that goes out, so settings can be adjusted to allow for individual preferences. Members are assigned roles with varying levels of permissions: member, manager, and owner. Multiple staff members can manage groups so that the unit is not dependent on one person to update membership or change permissions. Members can also be added so that they are able to send messages to the group, but do not receive any messages. Security settings allow the group owners to lock down the group as much or as little as necessary.

SITES

Google Sites is an application offered within the UCSC suite, but ILL has not yet explored what it can do for us. Sites allows multiple contributors to create content and upload files to a website. As with other Google applications, users can be given a range of permissions, depending on your needs. Google offers templates to give your site a more polished look. It does not require the creators to have any knowledge of HTML. Our circulation unit used Sites to migrate our physical procedures manual online. It is much more convenient to navigate to the Circulation Manual online and do a search for alumni, for example, which will search across all the manual web pages, instead of finding our clunky binder and flipping through the pages while someone waits for you across the circulation desk.

GOOGLE APPLICATIONS IN ACTION

Bringing new hires into the department proved to be an excellent way to test the usefulness of Google Applications. ILL hired two library assistants in summer 2012, a few months before the campus went live with Google Applications. Migrating department knowledge into shared documents is an excellent way to introduce new hires to procedures. It also cements the idea that everything you need to do your work, or almost everything, is available to you in the cloud. The project ILL undertook to update our automated notices was a great opportunity to model the participatory culture we espouse. Each ILL assistant was encouraged to add their own comments to the shared documents. Even if it is just to say "Yes! I like this bit!" it gets everyone into the groove of contributing. Over time, as new hires become more accustomed to the work and have more to contribute, the comments become more impactful.

We also had an opportunity to road test Google Applications in November 2013. A union strike left the campus inaccessible for a full day. We had advance notice of the action, and some staff members opted to work from home. Google Applications makes working from home seamless. Almost everything we need to do our work is now available in the cloud, thanks to Google Chrome bookmarks, in addition to the suite of applications. In the past, staff working from home would have to do a lot of planning and preparation in order to make sure everything they needed to do their jobs off-site was available. Now, there is very little that cannot be handled away from the office.

Google Applications are becoming increasingly ingrained in day-to-day library workflows. I began working on an ILL/resource sharing conference planning committee this winter—my first experience on this (or any) planning committee. I was thrilled to see that all the planning documents were stored in a shared file on Google Drive. I perused job descriptions for all the various planning committee roles, looked at planning documents from previous years, saw the breakdown of expenditures, and reviewed presentation proposals. For our first meeting, the agenda was created on Drive, and while there was a specifically identified note taker, everyone was free to add their thoughts to the notes. Thanks to all the documentation that was shared with me in advance, via Drive, I was able to get my footing quickly and easily.

GOOGLE TAKES US WHERE WE WANT TO GO

When I began working in interlibrary loan ten years ago, we had designated borrowing assistants and lending assistants. If a patron called with a question about their request, you had to find the borrowing assistant. If another library called with a question, you called on the lending assistant. And if one of those people was out for a day, or longer, patrons did not get the service they needed. Over time, we changed the way we do ILL work. We cross-trained staff members. We broke down the barriers between job duties. We reclassified our jobs, so that we could all do the same work without working outside of classification or not being compensated appropriately. We made significant cultural changes, among them how we made decisions and how we documented our work. This type of work had already happened in other libraries. ILL's transformation prompted other departments within the library to take similar steps. It was all part of a larger movement toward shared, collaborative work.

Creative approaches to work are commonly highlighted within big tech companies, Google among them. Google creates tools that allow multiple contributors to achieve a shared vision. While there is often a need for someone to own the work, at least as far as owning a shared document on Drive, a Site, or a Group, Google allows for the ability to designate privileges for as many people as desired. Everyone can pitch in to accomplish the goal. Google Applications can be used to move toward shared work and shared responsibility. At UCSC ILL, Google Applications allowed us to take a deeply held belief in the importance of communication and collaborative work and fully incorporate it into our daily operations.

Chapter Ten

Using Google Calendar Collaboratively for Library Organization

Misti Smith

Organization in a workplace is very important, and when it comes to a library, organization is really what a library is all about. Outside of clearly organized bookshelves, periodical displays, and video collections, the library is a space that is used for many purposes and is the center of a community, whether it's a public library or a higher education library that forms the heart of the college campus. Because the library must serve so many different patrons and groups, it is important that the staff is able to know what events are taking place and when and where. Knowing that this is an important need, when our college decided to implement Google as our mail system, which came with the entire Google Apps suite, it was a perfect opportunity to take advantage of the suite for organization. The specific app that was the most helpful right away was the Calendar with some help from Google Forms and Sheets.

LIBRARY BACKGROUND AND ORGANIZATIONAL NEEDS

The Mount Aloysius College Library sits atop the highest point of campus, which was previously the "top" of the campus, but now with some new buildings the library is more the center of the campus. Because of our central location, it is a popular spot for hosting events that are interdisciplinary and reach a wide variety of interests. The library hosts many faculty-centered events that focus on teaching and learning and also many student events that focus mainly on student success, including study habits, writing skills, and time management. Tutoring services are also provided in the library, which brings a large amount of traffic in and also increases the use of our various spaces. Like all libraries, the Mount Aloysius Library is always encouraging more usage of the library and all of the spaces as much as possible, so the problem of needing a system to organize the events is actually a good problem to have.

The library has a variety of spaces that have been purposed and repurposed over the years, but currently has the following main locations: the Practice Presentation Room, the Technology Commons, the Buhl Lab (a computer lab/classroom), and the Library Classroom. Other important spaces in the library that do not get reserved regularly are the Learning Commons, study rooms, quiet study area, and the Ecumenical Studies Collection. Because there are multiple spaces that are used by a variety of groups on campus it became important to devise a method to schedule these rooms and to do it in a way that keeps everyone informed about the scheduling so we do not have multiple bookings. Our previous calendaring system did not have a way to do this, so when Google arrived at our campus, I immediately got to work figuring out how we could best use it to manage a busy library.

STARTING OUT WITH GOOGLE APPS

In summer 2011, our campus Information Technologies Department made the choice to migrate from GroupWise to Gmail. The move was made for many reasons, including functionality and cost. While the move primarily affected the employees and students with regards to e-mail and calendar, many other tools are a part of the Google Apps suite that are highly useful in the educational setting. Realizing this, I got to work on completing the Google Certified Individual program to learn more about all that Google Apps has to offer. This knowledge was essential for creating and implementing campus-wide workshops so that our employees could take full advantage of the tools, while also allowing me to put them to use in my workplace, the library. The first office-wide feature I put into place is the sharing of our personal calendars, which was a first step in getting the rest of the staff more familiar with calendaring before I implemented the shared calendars and organizational scheme. Share settings in Google Calendar are accessed by clicking on the drop down next to the calendar name, in this case each person's own name, and selecting "Share this calendar." I required the staff to set their calendar share settings to "Share this calendar with others," and under that selection to share with everyone in the organization of Mount Aloysius College (see figure 10.1). Next to that option is a drop down that allows you to choose between seeing all event details or seeing only free/busy times. Staff were instructed to select the free/busy times option, which allows others on the staff, or in our case, at the entire college, to see when they are busy but not the details about what they are doing. This allows everyone to maintain privacy on their calendar while also giving others necessary information for scheduling meetings.

Once everyone had these options selected, we were able to more easily schedule staff meetings or small group meetings because the scheduler could quickly see when everyone had free time. Without choosing this option manually, Google will default to no sharing at all, which will not allow the scheduler to see anything on the other calendars.

SHARED CALENDARS FOR ORGANIZATION

After personal calendars were set up and shared properly, I moved on to creating shared calendars for the various functions that the library and its staff are a part of. The first one we addressed was the issue of multiple events by different groups using library spaces. We needed an easy way to schedule these spaces that could be seen and used by all staff and in some cases by students. There are two spaces that need to be able to be scheduled by students, faculty, and staff, the Technology Commons (ITC) and the Practice Presentation Room. There are two additional spaces that are only scheduled by library staff, the Buhl Lab and Library Classroom. The initial setup is the same for both types of calendars, but the difference lies in how they are used to schedule events. The following list describes the steps involved in the process.

- Step 1—Create a new calendar using the drop down next to "My calendars." At that point you need to enter a name for the calendar. A description is optional.
- Step 2—Decide who the calendar will be shared with and to what degree. The same share settings are available as shown earlier in figure 10.1. In

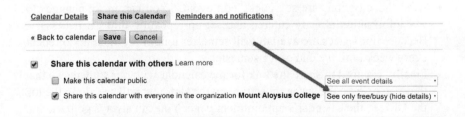

Figure 10.1.   Share settings on Google Calendar.

these cases, we do not need to share the calendar across the entire campus, since it is for the library staff only.

- Step 3—Share the calendar with specific individuals. Decide on what abilities they will have within the calendar.
- Step 4—Choose the specific settings. After entering the person's e-mail address, you can choose to allow them to "see all event details," "see only free/busy," "make changes to events," or to "make changes and manage sharing." You will need to add everyone with the ability to make changes, or you can choose "make changes and manage sharing" if you want everyone to have full owner rights to the calendar. The "make changes" option will allow everyone to see, add, or change events, but will not allow them to share the calendar with others. Four calendars were made using this same process representing each of the library spaces that needed scheduling options.

At this point, the Library Classroom and Buhl Lab calendar were complete because all of the staff had ability to see, add, or change events, and because of how the Google Calendar system is set up, the calendars can overlay one another, so it's very easy to see at a glance what space is being used and when. Each calendar can be assigned a different color of your choosing so that when they are all shown at the same time you can easily distinguish between the various calendars.

Next I had to make the ITC and the Practice Presentation calendars available for students to reserve. Because we do not want the students to have the same privileges that the staff do, we used an option in Google Calendars called "Appointment Slots."

- In the week view on the calendar, click a time to start the slots and then choose the option for "Appointment Slots" (figure 10.2). At that point, use the drop down to choose which calendar should have the appointment slots. The default will be your own personal calendar.
- Next choose how you want the slots set up, one single slot or slots of a certain length. For example, if you choose sixty-minute time slots between 1:00 p.m. and 5:00 p.m., you will end up with four available slots.
- After those options are set, click "edit event." You can make changes to any information already entered and set up repeating times if necessary. Because our spaces are available all semester long, I set them up to repeat every week until the end of the semester.
- Finally, we need to share the link for the calendar with our students so that they can reserve the space. In setting up the slot, you will see along the top the URL to the calendar's appointment page. You can always go back and find this by clicking on one of the slots and selecting "edit event." I placed the calendar's appointment page URL on our library web page.

STREAMLINING INFORMATION LITERACY SESSION SCHEDULING BY ADDING GOOGLE FORMS

For added ease of use, we combined the scheduling of our in-house information literacy sessions with the use of our shared calendars. A few years prior the previous library director had implemented an Information Literacy Plan that coincided with our Undergraduate Research Plan and designated certain courses, which aligned for the most part with years of study, to incorporate particular library information literacy sessions. For example, year 1 students in their freshman seminar would learn about plagiarism, year 2 students in their initial English course would learn about search strategies for searching the library resources, and so on throughout the students' senior year. In addition to these specific sessions, the library also offers other sessions that meet specific needs, such as specialized database instruction or iPad sessions, as well as unique sessions that can be requested by faculty to meet their own needs. Scheduling these sessions can get overwhelming and confusing when individual librarians are taking calls, e-mails, or visits from faculty and requesting spaces as well. We decided this process should be centralized to avoid confusion or overlap, and as an added bonus, it alleviates the need to collect quantitative data for these sessions.

The first step in this process was to create a form that could be used by faculty and librarians to request information literacy sessions. At the beginning of each semester, the link to the form is e-mailed out to all part-time and full-time faculty so that they can schedule these sessions. Google Forms can be accessed through Google Drive by clicking "create," then "form." You

Figure 10.2.   Setting up appointment slots.

can start by choosing a title and theme for the form and then clicking "ok." At this point, there are some settings at the top, one sample question, and confirmation page settings below. For form settings, we chose to require the user to log in to view the form and also to automatically collect the respondent's username. This allows us to obtain contact information for the person filling out the form without requiring them to type in their name and e-mail address. To start our form, we created a series of questions that gather basic information such as class time/day, course number, how many students are in the class, and so on. We also filled in a form description to give more information about how the library will use the information required by the form. After this basic information, we need to gather information about the level of course and the type of session the faculty are looking for.

Google Forms offers some more advanced options to tailor the form in a way that is more user friendly. In this example, we will use separate pages in the form so the respondent only has to see the information and answer the questions that are relevant to their request. Following the basic course information, I created a new page that has one question. The question asks what level the faculty person is teaching, and their answer to this question will route them to the section of the form that is pertinent. To do this, create the question as a multiple choice question, complete the question title, and list the choices. Next to the drop down for "Question Type," check the box that says "go to page based on answer" (see figure 10.3). Each answer will then need to be routed to its own page. Answer one can continue on to the next page, which is page three, answer two will go to page four, answer three to page five, and so on. Before you can select these pages, you will need to create them first. The pages can be blank at this point. Make this a required question so that respondents get the best experience with the form.

Each page on our form has one question that gives the respondent the options of information literacy sessions available for that particular level. We

Figure 10.3.   Creating a question that routes respondents to a particular page.

also give an option for "other" where the respondent can specify a unique session they would like. Each of these pages then reroutes the user to the final page of the form. On this page they are asked to add any additional comments or specifications they would like for their session. This question is set up as a "paragraph test" response rather than multiple choice like the others.

After completing the final question, the respondent is taken to the confirmation page of the form. At the bottom of the page during the edit form process, change the settings for the confirmation page. The default text reads, "Your response has been recorded." In our case, we edited that text to include a message telling the respondent to click on the link to access our calendar and reserve the time and day of their choosing. To make this possible I first had to set up a new calendar with appointment slots exactly as described previously. Because the appointment calendar URL is very long, I ran it through a URL shortener tool, in this case I used bit.ly. If the respondent is not comfortable with scheduling the appointment, it is explained that they will be contacted by a librarian who will ask them to choose a day and time.

While the form is able to integrate the scheduling into a streamlined process on the faculty end, we still have the issue of deciding which staff member is available to do each session. We addressed this issue by having the form responses go directly to our library secretary, who would be in charge of doling out the workload among us. At our library we have designated liaison areas, so the secretary's first choice is to assign the librarians to sessions that fall within their area. If that person is not available, she can then choose from the others based on who is available and who has the lightest instructional load. We also have many general sessions that could be done by any librarian, so the secretary makes an effort to keep the numbers even. After choosing an available librarian, a space and time is selected if the faculty member has not already scheduled one. Once the secretary has both pieces of information, she creates the appointment on that particular calendar, for example, the library classroom calendar, and invites the librarian and the faculty person to that meeting. The invite box is on the far right when editing an event. Because the calendar is tied to the e-mail system in Google Apps, the secretary simply has to begin typing the name of the person she would like to invite and then choose the correct person from the list. Because the librarians have set up their calendars to show free/busy times (see figure 10.1), the secretary can easily see if there is any conflict when she goes to invite them to the meeting. The librarian will then get the invite, and once they accept it, it will automatically be added to their personal calendar.

The final step in this process is choosing the destination for your data collected from the form. Every question that was set up in the form will be collected and displayed in Google Sheets, which is much like Excel. In the

edit form screen, along the top of the page select "Choose response destination." This will give you the option of creating a new sheet or adding the
sheet to an existing Google sheet. After you create this sheet, you will then
have in your Google Drive one file that is the form and one file that is the
responses to the form. By default the response sheet will be titled the same as
your form with "(Responses)" after the name. The responses file is where
you will go to see what your respondents have selected. In order to be aware
of new incoming form submissions, you will have to set up notification rules.
In the responses file, under the tools menu, there is a "notification rules"
option. There are a few different options, but the safest choice is to be
notified when "Any changes are made." This will alert you when a new
submission is made. You can also choose when you would like to be notified
in the rules settings. The choices are to be e-mailed right away or to receive a
daily digest. This decision is contingent upon how quickly you need to reply
to submissions.

CONCLUSION

Overall you can see how, with very little set up and training, Google Calendar, Forms, and Sheets can be used to organize your library spaces and your
staff. This system has worked very well for us, and I am always looking as
new tools and products become available as part of the Google line-up. Even
though our staff is quite varied as far as technology abilities and years on the
job, there was no member of the staff who was unable to pick up the system
quickly and implement it right away. As Google continues to grow and
partner with other products, I hope to keep this library on the cutting edge of
these tools.

Chapter Eleven

Using Google Sites as Library Intranet

Diana K. Wakimoto

Intranets have long been an important tool in libraries for document storage and communication. However, they are not without their problems and many are in need of updating with today's technology. While large libraries with robust, in-house technical staff are able to host and maintain their own intranets, smaller libraries are not, due to limited or no in-house technical staff. And even if you have a library intranet, sometimes navigating the platform is difficult, making it time and labor intensive to find documents, and decreasing the likelihood that staff members will use it. Library staff members need intranets that make use of today's online technologies in order to more effectively and efficiently do their work (Abram 2013, 26). Using Google Sites, even small libraries and libraries with limited information technology support are able to take advantage of a library intranet. Having used Google Sites as a library intranet for over two years, I can speak to its features and usefulness that have made it a good fit for our library.

GOOGLE SITES 101

Google Sites is part of the Google suite of applications (aka apps) that enable online, cloud-based, collaborative work. Google Sites is a website-creation platform that is free to use and does not require any coding knowledge, making it easily adaptable for library intranets. It is a great platform for multiple authors to use, especially for staff who are less comfortable with coding and website building (Jensen 2013, 121). If you can type a document on a computer, you can create a website! You just need a Google account to sign up for Google Sites, both of which are free. With more colleges and universities moving to use Google's suite of apps, many academic libraries are finding that it makes sense to use Google Sites.

One of the great advantages to using Google Sites as a library intranet is the fact that no software needs to be installed on local servers or computers. Everything is hosted by Google, and this negates the need for internal information technology support in installing or maintaining the intranet created using Google Sites. Also, because it is a cloud-based platform, Google Sites is accessible from wherever you have an Internet connection. The interface is browser based, which makes updating the intranet and accessing needed information easy for library staff.

Planning Your Google Site Intranet

There are two ways to get started with planning your library intranet using Google Sites: you can either dive in and create a site or plan out the basic structure and uses of the intranet first. Either way works; it just depends on how you learn best. If you prefer the "play first, then plan" approach with new tools and technologies, I suggest creating a test site. This will allow you to "kick the tires" of Google Sites, so to speak, and ensure that you feel comfortable with the editing features. By playing around with the Google Sites tools, you will probably also get an idea of what types of pages you need to create and what would be the most logical structure for the intranet based on the needs of the library staff. You can always delete the site when you are done testing and move on to creating your real intranet. And, because Google allows you to create multiple sites with one Google account, you can also leave your test site intact as a personal "sandbox" to test new layouts and features before implementing them in the library intranet.

The second way of going about building your site is to plan first and then play. If you already feel comfortable working in Google Sites, then you may want to first plan the structure of your library's intranet and then get to the actual building of the site. This planning will save you time from having to go back and restructure your site and works well if you have already worked with Google Sites and are comfortable with website creation. Either way, setting up the first part of your site will take only minutes, and you have all the flexibility to create the intranet that your library needs.

Setting Up Google Sites

Setting up a site is a simple process, from start to seeing the first page of your website takes less than five minutes. After logging in to your Google account, just select Google Sites from the list of apps, and after clicking the "Create" button, fill out the form to create your site. All you really have to do is name your site, type in a CAPTCHA, and your site will be created. You have many options for customizing your website and can even brand it with your library's name, logo, and colors. There are numerous templates that you can use to change the look of your site, and you can further customize these templates, along with creating numerous web pages that can act as generic web pages, file cabinets, lists, and announcement pages. The Google Sites help page is an excellent resource, full of clear, concise tips and trouble-shooting for adapting your website to use as your library intranet. Using the full range of tools available within Google Sites makes for a robust and effective library intranet.

After creating your site, and making sure the share settings are set to private so only library staff members can view the site, you will want to invite users. You can easily invite other library staff members to become part of your site by adding their e-mail address to the site and sending an e-invitation. This invitation will provide a link back to the site, allowing the staff members to log in and begin viewing and editing the content, depending on the level of permission you assign to them. The invitation process is straightforward and makes it easy to maintain an updated user list. Simply add or remove users via the share settings by adding or deleting e-mail addresses. This makes the administrative aspect of the intranet efficient, giving you more time for other tasks.

One of the features of Google Sites that will make it seem familiar to users is that the toolbars used for adding and editing content are very similar to those of any word-processing program. This makes it easy to learn Google Sites and work within the platform as you do not need to learn any coding or complex tagging system in order to add or edit content. Most library staff members should be able to start using the library intranet confidently after just a bit of exploring, without a long training course. And, as it is easy to revert to older versions of the website and "undo" mistakes, library staff members should be somewhat reassured that they cannot "break" the library intranet. Familiarity in document editing and layout serves as another reason to choose Google Sites as a library intranet platform. However, familiar and easy-to-use content editing tools should not be conflated with a limited or lacking set of customization and use options. On the contrary, Google Sites is very flexible and extensible for use as an intranet.

The flexibility of Google Sites in creating an intranet, or any website, is useful for branding, communication, and organization. Although Google Sites provides multiple templates and layouts for the entire website, you also have the ability to customize these templates and layouts. This allows library staff members to brand the intranet so it has a similar visual appearance to the rest of the library's online materials, as well as provide the flexibility for staff members to revise pages and make them into what they need to accomplish their work.

ADAPTING GOOGLE SITES AS A LIBRARY INTRANET

After setting up a site, it is easy to see why it is an effective and easy website platform. But what about using it as a library intranet? With an intranet, you are usually looking for three things in the platform: a private, online way of communicating among library staff members; an effective space for storing and sharing documents; and a way to track projects and project planning. Happily, Google Sites provides for all three of these aspects, making it a solid choice for use as an intranet.

Privacy and Permissions

Google Sites can be made private so that only people who are invited are able to view the site. This enables a created site to act as an internal intranet allowing for private communication among library staff members. Google Sites not only allows for creating whole websites that are either publicly or privately viewable, it also has page-level permissions, which will be discussed shortly. As noted previously, the first thing to do to create a private library intranet is to change the default Google Sites sharing setting from public to private. By making this universal change to the site, no one without permission can view the contents of the library's intranet, and the intranet is not discoverable via the open web. Only invited library staff members can view and edit the intranet. One caveat: although your intranet will be private, so that only those you invite will be able to access it, all information on the site is stored in the cloud. Although this storage is private, it has the potential to be accessed by hackers, and I would suggest not placing sensitive personnel information on the site. And, as always, follow all policies and guidelines at your institution to ensure compliance with data security protocols. Being smart about data security will ensure your data is safe and library staff members can effectively use the library's intranet.

One of the really powerful features of Google Sites is the ability to assign different levels of permissions to different people who have access to your site. This allows for more control over authoring content. The person who creates the site is automatically an owner.

Owners

Those with owner permissions are equivalent to the super admin accounts on many library software products. The owner can do everything from editing and adding pages to changing the colors and themes, and even deleting the entire site. You want to choose owners carefully, and it is often best to only have one or two owners for the site, if for no other reason than to limit the possibility of your intranet being accidently deleted.

Editors

Individuals with a step down in permissions are the editors of the site. Editors can add, edit, and delete pages; upload documents; and embed widgets. Editors do not have the same full range of administrative permissions as owners. For example, editors cannot change the theme of a site; however, both editors and owners can share the site with others and invite other staff members to become members of the site. Editor is probably the level most staff members will need to use and contribute to the site. Being an editor can also provide comfort to staff members who are not as comfortable with using intranets as there is no way for them to accidently delete the site or initiate a large change to the design or layout, such as changing the theme or colors of an entire site. This also allows the owners to maintain consistency throughout the intranet.

Viewers

Finally, there is the possibility of having viewers of the intranet. This permission level is for people who only need to be able to view the information without adding any content. Adding viewers to the intranet makes sense for volunteers and interns who need to have access to information, but who may not need to do any content editing in the course of their duties.

Page-Level Permissions

In addition to the overall, global privacy options offered by Google Sites, you also have the option of individual, page-level permissions. Page-level permissions allow for a more granular control over the intranet. This feature could be used for pages that only certain individuals or working groups need to see or use. These permissions can also be easily edited by the owners of the Google Sites intranet. Page-level permissions are especially useful if, for example, there are documents that interns or volunteers do not need access to or drafts of a working group that are not yet ready to be shared. Library staff members could also use page-level permissions to restrict access to home contact information or other information that needs to be easily accessible but with more controlled access than the rest of the site. All pages automatically have the same permission setting as the overall website; changing page-level permissions allows you to make a page with more restricted access. The help page on page-level permissions is especially useful and features examples if you are considering implementing this level of permissions.

Enabling Communication

All of these permission levels and privacy settings enable the library staff members to use the intranet to communicate and share information with those who need it. The granularity in permissions and settings allows communication to take place between people in different groups, control over web pages to be maintained, and information to be shared as needed. After setting privacy and permission settings as desired, you will be ready to upload documents to store and share on your intranet as detailed in the next section.

Document Storage and Sharing

One of the key advantages to using Google Sites as an intranet platform, as opposed to using a platform such as wiki, is the ability to easily upload documents to Google Sites and share them with others. This makes Google Sites very useful as an intranet, especially for sharing policy and procedure documents as well as meeting agendas and minutes. Each uploaded document is timestamped, making it easy to find the latest uploaded document and providing a way of tracking versions of the same document.

The file cabinet template for web pages makes it easy to both upload and share documents with a clean, organized interface. This organization capacity, including the ability to create folders, is especially important when uploading a large number of documents. In addition to creating folders to help organize documents in your online file cabinet, you can also write short, descriptive notes that display next to the filename of your document. This feature is a great way to ensure that people find the correct document before downloading it.

In addition to simply uploading documents to Google Sites, there is also the opportunity for integration with Google Drive. Google Sites allows both embedding and linking to documents in Google Drive, providing another avenue of online collaboration via the library's intranet. This ability to collaborate in real time on documents online is an advantage both in communication and in project management, which is the issue of the next section.

Project Management

In addition to being able to communicate and easily share documents, using Google Sites as an intranet allows you to coordinate projects and teams. Taking advantage of the cloud storage of Google Sites, project managers and team leaders can use Google Sites to effectively organize projects online. Using the announcement web page template allows for easy communication of project deadlines, reports, and meetings. Intranet users can even subscribe to announcements and page revisions so they receive an e-mail alert when pages that affect their projects are updated. This is a time-saving feature, as staff members do not have to check the intranet to determine if anything has been updated, but can instead be "pushed" announcements and notifications.

Revision history can also help in project management, as Google Sites automatically records who has edited the various pages and when. This can serve two purposes: (1) to determine who has been working on the project and (2) as a failsafe in case something happens during a revision that is unintentional. A potential difficulty with any team or group project work is determining if all parties involved are actually working on the project. By seeing who has added to the site and what they have contributed, revision history acts as an incentive. Everyone can see who has done what. Also, as accidents do happen, the ability to reload an earlier version of a web page from the revision history is a valuable safety net.

Two other embeddable widgets help make using Google Sites for project management very effective: Countdown Clock and Google Calendar. You can embed a customizable countdown clock into your project pages, which helps with determining at a glance how much more time you have to work on a project. The Google Calendar widget is useful for sharing time lines and deadlines. After creating a calendar for your team, you can embed it in one of your team's pages on the intranet, ensuring everyone is aware of upcoming deadlines.

GOOGLE SITES USE AT THE CALIFORNIA STATE UNIVERSITY, EAST BAY, UNIVERSITY LIBRARIES

The librarians at the California State University, East Bay, University Libraries currently use Google Sites to serve as the librarians' intranet. Because Cal State East Bay employees have access to Google Sites as part of the university's Google Apps suite, it made sense to turn to Google Sites as a platform. We use a main site to share documents, agendas, minutes, and other important information such as committee assignments. We have additional sites created for collection development, exhibits, and instruction. Together, these sites have allowed for effective online collaboration, project management, and efficient document storage. Also, due to the easy-to-update nature of the platform, we have been able to keep the information current and instantaneously share new information.

Prior to using Google Sites, we used a wiki as an intranet; however, this situation was not ideal for our needs. While the wiki was private and we had the ability to add and edit content on the wiki, many librarians found the wiki markup language cumbersome, and often the content did not end up formatted as expected. Furthermore, we did not have the ability to upload documents to the wiki, which greatly limited its usefulness as an effective intranet. We were eager to try out a new platform when Google Sites became available. While it took a few days to migrate all the needed information from the wiki to the Google site, it has been a successful transition to using

Google Sites as the new intranet. The flexibility and extensibility of Google Sites has made it a good fit for us at Cal State East Bay.

CONCLUSION

Library intranets are important tools, and Google Sites provides a useful platform option for many libraries. With its ease of use, flexibility, and no need for extensive in-house information technology support, Google Sites is a platform that library staff may want to review when considering options for a library intranet. The ability to collaborate from anywhere with an Internet connection is a definite benefit, as well as integration with other Google Apps and third-party apps, which makes for a very extensible tool. In times of limited budgets, having a free intranet solution is a very appealing notion, and Google Sites is both free to use and useful, the best of both worlds.

REFERENCES

Abram, Stephen. 2013. "Next-Generation Intranets." *Information Outlook* 17 (5): 26–27, 29.
Jensen, Karen. 2013. "Managing Library Electronic Resources Using Google Sites." *Journal of Electronic Resources Librarianship* 25: 115–23. doi:10.1080/1941126X.2013.785289.

Part III

Library Administration

Chapter Twelve

Google Analytics and Library Websites

Wei Fang

Google acquired a web analytics firm called Urchin Software in March 2005. In November of the same year, Google released its online version of a website tracking service, named Google Analytics (GA), which has been widely used by companies and organizations worldwide since then, and is now the most widely used website statistics service (W3Techs 2014). Unlike the original Urchin, which was downloadable and priced from $899 to $4,995, GA-hosted service is provided without charge for up to 10 million hits per month per account. If your site exceeds that limit, Google also offers GA premium service for an annual flat fee of $100,000.

Google has continuously redesigned its GA user interface (UI) in order to give its users better use experiences while providing more functionality. The latest redesign was released in 2013. In addition to UI, GA also released new tracking libraries in the form of JavaScript snippets to gather deeper analytical data. GA started with tracking libraries named urchin.js and ga.js, and the latest library is analytics.js. All tracking libraries are hosted by Google, and it may update the codes without informing end users. In this chapter, I use the newest interface and functionalities of GA. In particular, I will discuss some essential GA functions useful to libraries.

WEB ANALYTICS AND GA

Web Analytics and How It Is Measured

According to the web analytics definition by Digital Analytics Association (Web Analytics Association 2008), "Web Analytics is the measurement, collection, analysis and reporting of Internet data for the purposes of understanding and optimizing Web usage."

Web analytics involves many terms, definitions, and methodologies. GA uses two key concepts throughout its web interfaces: dimensions and metrics. By combining different dimensions with metrics, GA can create a wide range of web analytics reports.

Dimensions

Dimensions are used to describe data as measurable characteristics or attributes of an object. For instance, a visit to a website can have dimensions such as session duration, entry page, exit page, and so forth.

Metrics

Metrics are used to measure data. If an element of a dimension can be measured as a sum or a ratio, it is called a metric of the dimension. For example, in the example above, average session duration could be a metric of the session duration dimension.

In order to have basic measurements about a website's traffic, there are four major metrics that need to be evaluated: page views, visits, unique visitors, and event.

Page Views   Page views refer to the number of times a web page was viewed by the visitors.

Visits   Visits, also known as sessions, record interaction between visitors and the website within a specified time period. When a visitor no longer interacts with the website within the specified time period, the visit is timed out.

Unique Visitors   After identifying and filtering out visits from web crawlers, such as spiders and robots, individual visits are counted only once in the reporting period.

Event   Event shows activities, such as viewing Adobe Flash content or changing form fields, that happened within a web page at a specific time and date.

With these four basic metrics, you receive information on how many pages were viewed through how many visits of how many unique visitors and what they did on the page.

Why Web Analytics and Why GA

Libraries have offered their websites, usually including main websites, catalog sites, interlibrary loan sites, digital repository sites, and proxy server sites, to their patrons since the dawn of the Internet. These websites have since become the bridge between libraries and patrons who seek accurate and accountable information. However, only a few libraries were able to analyze their websites' traffic data effectively, let alone understand how patrons behaved and why. Before GA, libraries used to gather subjective patron usage data via paper or online-based surveys. There were many problems with this method, mainly due to its subjective nature. Conclusions based on this subjective method may not represent what really goes on, and it may even be misleading. Web analytics can be performed based on server log files. Libraries that had enough resources performed their own server log analysis. There are many utilities and methods to analyze the same set of tracking logs. For instance, you can analyze Apache web server's log files using open source software like Webalizer to gain some insights about your site's activities. However, this method requires the server to be configured correctly so that log files can be recorded in a meaningful format and later be analyzed using analytical utilities. This was not changed until GA came along. Obviously, GA offers a more powerful set of functions with little code snippet.

TRACKING WEBSITE ACTIVITIES WITH GA

How GA Works in General

In order to have websites tracked by GA, a small snippet of JavaScript code has to be implemented immediately before the </head> tag in each and every page that needs to be tracked. When a page with a tracking code is requested by the website's users, the JavaScript code will be executed to call scripts that are hosted on Google's servers. Upon successfully executing the scripts, a wide array of activity data, such as the page the user is reading, and user end information, such as the user's operating system, will be submitted to Google and stored under your GA account. When a data analysis task is being performed from GA's web interface by an authorized user, the stored data will be analyzed by GA based on the configurations and criteria set by the user. The data will then be visualized and displayed.

Using GA as the Main Method of Library Web Analytics

Although there are many methods and utilities to analyze the same set of server logs or web traffic data, GA offers a powerful set of functions by default and makes them easy to use. GA takes all tracking data off the local server and stores it on its own servers. GA offers a real-time activity report that is perfect for monitoring the site when a major event is occurring. Most important, GA users can generate easy-to-read web analytical reports with just a few mouse clicks.

Because each and every data analysis performed by GA is based on analyzing web traffic data, GA reports are objective, and there is no room for guess work. When armed with these intuitive and objective web analytical reports, library IT staff or web services librarians have a complete and clear picture of how patrons use the site and what is really needed to optimize the site. It is also easier for the library management team to make decisions about how to provide a better web service to the websites' users.

In order to use GA to track a website, you must have permissions and privileges to modify your site's file directory and files hosted. You also need an active Google account before signing up for GA. If you don't have an active Google account yet, it is time to get one. Open up your web browser and redirect it to https://accounts.google.com/ . Then click on the "Create an account" link at the bottom and follow the instructions. Once you have created a Google account, go to http://analytics.google.com/ and log in with your Google account credentials.

There are three steps involved to obtain necessary tracking code snippets for your websites: (1) Have an active Google account and sign up for GA. (2) As soon as you sign in, click Admin in the menu bar, and create Account and Property based on your needs. For instance, you can create an account called "Library Web Tracking" and nest a property named "Main Library Website" under the account. Obtain unique tracking ID and codes for each website (or property, in Google's term) you want to track. After all the properties are created, select the property you are working with. Click Tracking Info/Tracking Code. (3) Locate the tracking code snippet that starts with <script> and ends with </script>. Insert the tracking code into the main landing page of your website so that GA can verify that you are the owner of the site and have the necessary privileges to modify web pages.

Using GA to Track Static or Dynamic Websites

From the main library website to proxy server site, almost all library websites can be tracked using GA. Once you have successfully obtained the unique tracking ID and appropriate JavaScript code snippet, you can start to insert them into your web pages or web page templates. It is absolutely critical not to mix up tracking code snippets from different properties or the tracking data will end up invalid and misleading. If you are working with static pages, copy the JavaScript code snippet that starts with <script> and ends with </script> from the GA admin interface and paste it right before the </head> tag in each and every page you want to track. For dynamic web pages, you only need to paste the tracking code right before the </head> tag in the template page.

In some cases, template pages need to be recompiled so that the new code can take effect. Take the popular digital repository system DSpace as an example. In order to insert tracking code into its JSP UI template page and make it effective, you need to find the build version template directory, such as /dspace/target/dspace-1.7.2-build.dir/webapps/jspui/layout, and modify the header-default.jsp by inserting the tracking code before the </head>

tag. After the modification is done, you need to recompile the WAR file for Tomcat.

Using GA with Drupal and WordPress

If you are using a content management system (CMS), your sites can be tracked as well. Most libraries are using Drupal or WordPress as their CMS. There are two ways that you can implement the GA tracking snippet in your Drupal- or WordPress-driven sites. One way is to embed the tracking snippet into your site's templates right before the </head> tag. The other way is to use the Drupal Google Analytics module or WordPress's Google Analytics plugin. The second option is recommended since it will cache GA code on the local server to improve loading time and CMS users have the ability to exclude certain users, roles, and pages from GA tracking.

Install and Configure the Drupal GA Module (https://www.drupal.org/ project/google_analytics)

The installation process is simple: download the module, unzip it to a local directory, and upload unzipped files to your server's /sites/<yoursitename>/modules/. You will then log in to your site with an admin role, which allows you to add this module to your site via the menu Administer, Modules. Locate the checkbox next to Google Analytics and enable it. Save the configuration. To configure the module, go to Administer, Site Configuration, Google Analytics, enter the UA number GA assigned to your site, and make changes on the configuration page based on your needs. You need to click on the Save button once you make all the changes.

Install and Configure WordPress's GA plugin (https://wordpress.org/ plugins/google-analytics-for-wordpress/)

To use the WordPress GA plugin, you need to download it to your local machine, unzip it to a local directory, and upload unzipped files to your server's /wp-content/plugins/. You will then log in to your site with an admin role, which allows you to activate this plugin through the Plugins menu. To configure the plugin, go to Options, Settings, enter the UA number GA assigned to your site, and make changes on the configuration page. You need to click on the Save button once you make all the changes.

USE GA IN LIBRARY REPORTS

You can start tracking your websites as soon as JavaScript code snippets are installed in all the properties that you want to track. However, you can create meaningful GA reports only after you have accumulated enough data. It is important to understand that you must have one or more time ranges for every report that GA creates. For instance, you can ask GA to create reports for the past day, week, month, or year. You can also set the time range to a specific starting and ending date.

Administrative Features Useful to Libraries

Websites administers can take advantage of the following GA features: change history, intelligence events, filtering, and comparing reports between two time ranges.

Change History

This feature offers a list of user activities in your GA account. It provides a snapshot of who did what and when to your GA account. For example, if you have multiple users accessing your library's GA account to manage different properties, this feature records all user activities. It shows records such as at a specific time User One was granted permission to read and analyze Property A. This is a useful feature if there are security concerns or if some unwanted changes were made unintentionally. To access this feature, click Admin at the top menu bar and click Change History at the left-side column under Account Section.

Intelligence Events

Intelligence events is a group of alerts that GA sets up to alert its users of the performance of a key metric in a segment within a specific time period. It is similar to the physiologic status monitoring system used in emergency rooms. By default, GA offers a set of automatic alerts that cover key metrics of a website. The total number of automatic alerts that GA creates is based on the activities and uniqueness of the website. For instance, when GA senses that most of the website users came from a certain state, an automatic alert will be created to trigger if the total sessions from that state increases or decreases. GA will also automatically assign an importance indicator to each alert it creates. Users can set up custom alerts as well. You can configure e-mails or text messages to be sent out if an alert is triggered. For instance, you can create a custom alert to let you know when all your websites, or just one of your websites, has less than a certain number of new user visits in a day.

Filtering

Filtering is a very powerful feature. With this feature data collected by GA can be filtered, and only the filtered data will be used to create reports. For instance, you can set up a filter to exclude all traffic from a domain or IP address. This can be used to exclude all internal access generated by library staff members so that all the reports show only patrons' activities. You can also create custom filters to filter out metrics that are not of interest. For example, a custom filter could be set up to exclude certain pages from the total page view count. Filters can be set up to include specific data as well. Configure filters by clicking Admin in the top menu and then the Filters link under the View tab on the right side. It is important to adjust filters based on your needs. When a filter is no longer needed, it should be removed.

Comparing Reports between Two Time Ranges

In GA, data can be summarized or grouped by day, week, month, and year. You can also set up two time ranges and compare the results. At the top-right side there is a time-range indicator. Click the drop-down box and check the "Compare to" box before clicking the Apply button. This will allow you to compare two time ranges. This is useful when you are trying to make sense of the differences between current state and past state. For instance, you can use this feature to figure out if the website is seeing more visitors compared with the past. Reports such as this can certainly be used in the library's annual report. The time ranges should be set up in a way that makes sense to you. For instance, if you want to see last year's traffic compared to this year's traffic, you can do so.

Data Analysis Functions Useful to Libraries

Librarians can take advantage of the following data analysis functions: dashboard (visualized summaries), pages (content by title), in-page analytic, traffic sources overview, behavior flow, content drilldown (e.g., visitor segmentation), data export, and e-mail and add other users. I will discuss these features in more detail below. It is important to know that I will not provide navigational details to the functions discussed in this section. Instead, you should take advantage of the "Find reports & more" function at the top of the left-side column under the Reporting menu. Just click Reporting in the top menu and type in the keywords used below. GA will lead the way.

Dashboard (Visualized Summaries)

This is the most useful feature when monitoring site activities. It is literally a nutshell of the website log information. The dashboard offers a visual summary of the website activities in a time range that you can customize. For instance, for a given time period, you can determine how many users visited the website, how many pages were viewed, how long on average a patron stayed on the website, which content was the most popular, and which website the patrons navigated from before visiting your site. When GA is set up, one dashboard is created automatically to include default dimensions and metrics. Multiple dashboards can be created to include different segments or dimensions.

Pages (Content by Title)

Pages present a list of the most popular items on the website. It is literally a top hit list of the content of your website. In the pages report, by default, web pages are listed based on the number of page views they get. Web pages are showed in the page column with their relative directories. This can be sorted by page title as well as by clicking at the Page Title link in the Primary Dimension menu bar.

In-Page Analytics

This feature used to be called site overlay. In-page analytics displays clicking summaries in small pop-up windows on an image of the actual web page. Figure 12.1 shows an example page; figures 12.2 and 12.3 show enlarged portions of the screen where the pop-ups can be seen clearly. In figure 12.3, for example, we see that the quick link to the legal databases has been used by 5.5 percent of online users. This feature gives us a straightforward way to find out how many users clicked on the links on that page. If the website redesign team is trying to figure out what the most popular links are and possibly why they are popular, in-page analytics serve the same function as a CT scan.

Figure 12.1.   In-page analytics.

Traffic Sources Overview

This feature summarizes the methods that visitors used to find the website. For instance, it shows how many users used a direct link to visit the website

Figure 12.2. Blow-up of News & Announcements section.

and how many users were redirected to the site via another website. With this feature, search engines and external websites that redirect users to your website can be easily identified.

Behavior Flow

Behavior flow (figure 12.4) offers a visualized user pathway by showing interactions between users and your website. It shows the navigational path of the visitors: the first page the visitors started with, then which pages they subsequently viewed, and eventually the last page viewed during their visit.

Content Drilldown

This feature shows how people find and interact with your web pages. By employing this feature, you can combine any GA report with other dimensions or metrics, such as country, region, and keyword, to generate a new report that presents detailed information regarding your visitors. For instance, you can see detailed information about visitors who viewed a certain web page and where they came from—that is, visitor segmentation based on region. This would be very useful when redesigning a library collections' landing page. Take figure 12.5, for instance; it shows that during the time

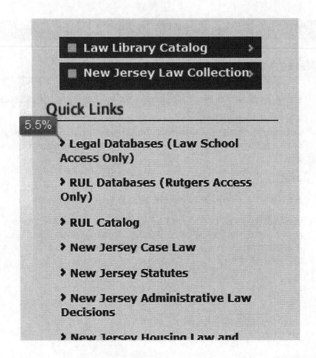

Figure 12.3. Blow-up of Quick Links section.

range selected, the Courts collection attracted more patrons than the NJ Status collection. Most people who were interested in the Courts collection were from Trenton, New Jersey. By changing the secondary dimension, you can find out the kinds of network connections visitors have, what web browsers they used, what operating system they used, and so forth.

Data Export

The GA data exporting tool and its API allow users to export report data in PDF, XML, CSV, CSV for Excel, and TSV. This feature is useful because it generates data that can be analyzed with other statistical programs. For instance, outputted data can be imported to Microsoft Excel or other statistical software for further analysis. When you are satisfied with a report, click the Report tab at the center top and select the formats you prefer. It is important to remember that the export function will only export the data that are displayed in your current report. If, for example, you want to export the entire data set of your website, you will have to use GA API. Please refer to the GA API site (https://developers.google.com/analytics/); I will not discuss GA API in this chapter.

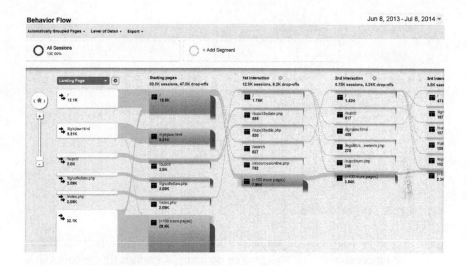

Figure 12.4.   Behavior flow.

E-mail and Add Other Users

By changing the settings of the GA service, an admin user can decide if adding or deleting users is needed. A user must have a Google Gmail account in order to be added to this service. Each user can be assigned different privileges in order to view various reports and profiles. GA also allows admin users to define a list of e-mails and reports to be sent immediately or on a scheduled basis.

Limitations of GA

1. The website that you are tracking cannot have more than ten million hits per month or GA will not be free.
2. Privacy can be an issue. As GA states on its website, it generates "aggregated non-personal information" to share with third parties. Thus, it is not recommended to use the GA service with high-security websites or when privacy is a major concern. Also, you may want to post a privacy statement about using GA in your library policy web page.
3. If GA is not configured correctly, the traffic data and reports will be invalid. If you are using the old ga.js tracking code, you should upgrade it to analytics.js.
4. If GA service is down temporarily, so are your website and tracking data. In this case, GA's tracking snippet would slow down the load

Page path level 2	Pageviews	Unique Pageviews	Avg. Time on Page	Bounce Rate	% Exit
	297,754 % of Total: 99.90% (331,012)	144,674 % of Total: 97.20% (148,821)	00:02:12 Site Avg: 00:02:07 (3.74%)	51.75% Site Avg: 51.57% (0.35%)	38.67% Site Avg: 36.14% (7.00%)
1. /courts/	143,788 (48.29%)	67,219 (46.46%)	00:02:25	55.27%	43.14%
2. /njstats/	67,090 (22.90%)	35,249 (24.36%)	00:01:37	27.22%	23.15%
3. /oal/	48,390 (16.25%)	22,027 (15.23%)	00:02:46	54.64%	41.95%
4. /ethics/	22,471 (7.55%)	13,214 (9.13%)	00:02:07	61.68%	50.52%
5. /drb/	14,049 (4.72%)	6,197 (4.28%)	00:02:01	48.52%	38.52%
6. /fed/	1,755 (0.59%)	653 (0.45%)	00:01:12	27.93%	25.19%
7. /njconstr	88 (0.03%)	52 (0.04%)	00:02:30	55.56%	37.21%
8. /munarb/	68 (0.02%)	20 (0.01%)	00:02:34	46.15%	22.06%
9. /gdoc/	51 (0.02%)	37 (0.02%)	00:02:48	70.00%	64.71%
10. /njleg/	5 (0.00%)	5 (0.00%)	00:11:51	100.00%	80.00%

Figure 12.5. Content drilldown.

speed of your website at the user's end. It may even cause your site to be inaccessible. This was proven on many occasions throughout GA's history. For instance, in the early dawn of GA, due to the popularity of the service, Google placed new applicants on a waiting list until GA became generally available to the public in mid-August 2006. This may not happen again, but temporary service outages may occur.

5. Certain content, such as XML and RSS feeds, will not be tracked accurately due to the fact that the tracking snippet will not be executed upon loading.

6. GA keeps updating its tracking libraries. In order to use up-to-date features, your sites' tracking snippets need to be updated as well.

CONCLUSION

Web analytics reports work like the GPS in our cars. Instead of aimlessly redesigning a website or improving certain services on the website, web analytics offers a way to make sense of a site's traffic data web. GA makes a complex job much easier with a powerful set of functions and an easy-to-use interface. When GA's web analytics reports are being used and interpreted correctly, libraries can use this wonderful service to optimize their website and services.

REFERENCES

Web Analytics Association. 2008. *Web Analytics Definitions 20080922 for Public Comment*. Web Analytics Association.

W3Techs. 2014. *Usage of Traffic Analysis Tools for Websites*. W3Techs. Retrieved April 10. http://w3techs.com/technologies/overview/traffic_analysis/all.

Chapter Thirteen

Google Video and YouTube

Deborah Hamilton

Google offers two video services: YouTube and Google Video. Google Video is a search tool to find video content on the web. It is particularly useful because, while YouTube searches only yield results from YouTube's website, Google Video will find videos from all over the web, including from YouTube. Google Video also allows for advanced searches that limit the number of results by using filters such as video duration, time of posting, source, and quality. Alternatively, YouTube has a much wider array of functions. In addition to searching YouTube for video content, users can upload and embed videos, create playlists, and subscribe to video channels. Both services can be useful to libraries and their patrons in accessing video content on the web. However, YouTube, by allowing library staff and patrons to sift, save, and share videos, can be particularly useful in the following ways:

- Create public relations videos to highlight library services, give virtual tours, create book trailers, provide collection spotlights, initiate support campaigns, and raise awareness of library issues (Purcell 2013, 14–16; Dankowski 2013, 38–41). YouTube is an inexpensive way to create an online presence. Using the embed feature, you can place videos directly on your website.
- Create tutorials to share with library staff and patrons; or find tutorials that will help both staff and users. Viewing other tutorials can be helpful in creating your own, and many tutorials can be repurposed for your needs.
- Use YouTube for technical support. There are numerous videos on YouTube that demonstrate how to use different types of software and equipment.
- Use the playlist feature to gather useful videos into one place. Or create playlists to collect videos created by your parent institution.

GOOGLE VIDEO

Google Video is a great way to search for video content from all over the web. You can access Google Video by doing a Google web search for Google Video. A basic search in Google Video is done using the search bar at the top of the page. After you have completed a basic search, look for the "Search Tools" link below the search box to apply search filters. Or if "Search Tools" is not visible, click on the gear icon to open "Advanced Search." Using "Search Tools" you can apply one or more of the following filters:

- Duration
- Time posted
- Quality of video
- Video feature (e.g., closed captioning)
- Video sources

You can use these filters to refine your search and reduce the number of results you would otherwise have to page through.

YOUTUBE

The video-sharing service YouTube was purchased by Google in 2006. YouTube allows users to both view and post videos. So, while you can search for and watch videos without logging in to Google, in order to use all of YouTube's features you will need a Google account. In this section, we will cover everything you need to know to utilize all of YouTube's features.

- How to create a Google account
- How to navigate and search YouTube
- How to create playlists and follow other users/channels
- How to upload videos to YouTube

How to Create a Google Account

It is necessary to have a Google account to access all of the features of YouTube. If you have a Gmail or Google+ account, this login will work for YouTube as well. If you need to create an account, go to the YouTube home page. In the top-right corner, click the "Sign In" link. On the next page, click on "Create an Account." Fill out the form on the following page. When you create your account, you can use an existing e-mail address or you can request a new e-mail address from Gmail.

After logging in to your account, you can adjust the settings of your YouTube account using the menu that is opened by clicking on the gear icon in the top-right portion of your screen. In the "YouTube settings" menu, you can adjust various settings:

- Privacy
- Basic account features
- Account connections
- Playback options
- Connected TVs

The "Help Center" is found in the same drop-down menu accessed by clicking on the gear icon. You can use the search bar to look for a particular topic, scan through their suggested articles, or click on the link for the Help Center at the bottom. The Help Center will open in a new web page. There you will find the help topics, including instructional videos, divided into categories.

How to Navigate and Search YouTube

When you arrive on the home page for YouTube, you will see that the screen is divided into three areas: the main frame, the top banner, and the left banner.

The Main Frame

When you first visit YouTube, this frame will feature YouTube's recommended videos. If you are logged in to your Google account, these videos are determined by your search history, subscriptions, and liked videos. After completing a search using the search bar at the top of the screen, the main frame will show your search results. Once you have selected a video to watch, it will appear in the main frame along with the video's information, comments from other users, and other video suggestions on the right.

The Top Banner

In the top banner you will find the following tools from left to right:

- The YouTube icon, which when clicked will always bring you back to YouTube's home page
- An icon with three horizontal lines that will open and collapse the left banner
- A search box
- The upload button

- A gear icon, which will only appear when you are logged in to your Google account and which will allow you to access your account dashboard, video manager, analytics, YouTube settings, and the help menu
- A login button that will allow you to access your Google account; or, if you are already logged in, this icon accesses your Google account

The Left Banner

Please note that the left banner of the screen may be collapsed depending on the size of your screen. You can open and close the left banner by clicking on the icon of three horizontal lines that is next to the YouTube logo in the top left of the screen. When the left banner is open, you will find links to the following:

- What to Watch: YouTube's video recommendations based on your viewing history.
- My Channel: Any videos you upload will be stored here. You can also access your playlists from this page.
- My Subscriptions: Access all the activity from the channels you have subscribed to.
- History: View the videos you have previously watched.
- Watch Later: Videos that you have marked to "watch later" will be saved here. In order to mark a video to watch later, look for the clock icon on the video player.
- Playlists—Favorite Videos: Keep track of your favorite videos. There is a link to add videos to this list on the Favorite Videos page. You can add videos that you have uploaded as well as videos created by others.
- Playlists—Liked Videos: Keep track of all videos that you mark as "liked." To mark a video as liked, look for the thumbs-up symbol on the video page.
- Subscriptions: Lists all the channels you subscribe to so you can access them quickly.
- Browse Channels: Browse the content on YouTube.
- Manage Subscriptions: Set your preferences for how you want your subscriptions displayed and updated. You can delete any of your subscriptions here.

Searching and Browsing YouTube

The search box appears at the top of every page in YouTube. To search, type your search terms into this box and hit enter. After you have entered your search terms and your results appear, you can add filters to your search. Look for the box labeled "filters" above your search results. Once clicked, a drop-down menu will appear with the various filters you can apply. These filters include:

- Upload date
- Result type
- Duration
- Features
- Sort-by method

It is possible to apply multiple filters to the same search. After you select a filter, you will have to click on the filter box again to reopen the drop-down menu. You can remove any of the filters you have applied by clicking the "X" symbol next to it.

You can also find content by browsing YouTube. The "What to Watch" page will show you what is popular on YouTube and will make suggestions based on your account's history. Or you can browse YouTube channels using the "Browse Channels" link toward the bottom of the left banner. On the "Browse Channels" page, you can either scroll through the list of channels or use the search bar at the top. When scrolling through the list, make sure to notice that the square icons directly below the category name each point to a different channel.

Watching Videos on YouTube: The Video Player

Once you have found a video you want to watch, click on the video's title. Along the bottom of the video player you will find icons to access the following:

- The play or triangle symbol will play the video. Once the video is playing, the double vertical line symbol will pause the video.
- The volume control is the speaker icon to the right of the play symbol. Hover your cursor over the icon for the volume control slider to appear. You can adjust the volume using this slider. To mute and unmute the video, click on the speaker icon.
- Next to the volume control, the video's time displays. You can see the amount of time you have watched and the total time of the video separated by the slash character.
- Click on the clock symbol to watch this video later. You can find the saved video by clicking on the "Watch Later" link in the left banner.
- Click on the box with "CC" to enable closed captioning. This is a great feature for users with hearing difficulties or if the video has poor audio quality.

- The gear icon will allow you to turn on or off any video annotations. Annotations are pop-up messages that may appear in some videos. The gear icon will also allow you to select the quality of video you are watching. If you find the video is having difficulty loading or playing, it may help to switch to a lower-quality version.
- The square symbol will open the video in the large player.
- The bracket symbol will open the video in full screen. Hit the escape key to leave full-screen mode.

Viewing the Video Information

Underneath the video, you will find information about the video and the user who uploaded it. Directly below the video is the title. Below the title, you will find the name of the user who uploaded the video. You can click on a user's name to access all videos uploaded by that user. Next to the user's name, you will see how many videos he or she has uploaded as well as the link to subscribe to that user's channel. A channel is a user's page where you will find all videos and playlists created by that user. Underneath the "Subscribe" link, you will see a thumbs-up icon and a thumbs-down icon. If you like a video, you can click on the thumbs-up symbol. If you do not like a video, click on the thumbs-down. To the right, you will see a tally of how many likes and dislikes a video has accumulated. The number above that tally is the total number of times that video has been viewed. Below the number of views, likes, and dislikes, you will see several links that allow you to access more information about the video and to perform several operations. Use the "About" link to find information about when the video was published on YouTube, what type of license the video has, what category it falls into, and any other information the user wishes to share. You can open the full "About" page by clicking on "Show More."

For some videos, you will see a document icon below the video. Certain videos have an available transcript; you can access it by clicking on this icon. The transcript will appear below with time stamps to show where in the video the dialogue occurs. The graph icon will allow you to access the video's statistics. These statistics track views, subscriptions, and shares over time. The flag icon can be used if you wish to report a video for inappropriate content.

Sharing Others' Videos

Under the video player, next to "About" is the "Share" link. Clicking on "Share" will open a submenu with three options: "Share this video," "Embed," and "Email." Under "Share this video" are the icons for several different social networking sites. Click on the logo of the site where you wish to share the video. Below the social networking logos, you will find a perma-link to the video. This link is always the most direct link to a video, and it is less likely to break than if you were to cut and paste the link from your browser's address bar. If you use the permalink to share a video, it is also possible to start the video at a specific time. To do this, make sure you check the box next to "Start at," then add the minute and second marker where you want the video to begin, and then cut and paste the link.

Under the "Embed" link, you can access the embed code. An embed code is a snippet of HTML code that you can add to a web page or blog that will allow the recipient to watch the YouTube video directly on the site where you place it, rather than being directed back to YouTube. Below the box with the embed code, you can select the size that you want the embedded video to appear. You also have options to display suggested videos after the embed-ded video finishes playing, to use the enhanced privacy mode, or to use the "old embed code," which functions better with certain website and blog platforms. Finally, you can share by e-mail. When you click on the "Email" link, enter the e-mail address of each person with whom you wish to share the video.

How to Create Playlists and Follow Other Channels

Creating Playlists

In order to save or return to videos, you can create playlists or subscribe to video channels. On a video's page, next to the main share link, you will see the option to "Add to." Clicking this will allow you to add this video to any of your playlists. If you do not have any playlists created, you will be given the option to create a new playlist under "Add to."

Or you can go to the "Playlist" link in the collapsible left banner. Once you are on the playlist page, click "New Playlist." Give the playlist a name, and click on "Create." To add a video, click on "Add Video." You can either conduct a search to find a video, enter the URL for a known video, or add a video that you have uploaded. To adjust the playlist settings, click on the box with the gear icon labeled "Playlist Settings." There you can adjust the priva-cy settings. The default setting is always public. Here is a breakdown of the privacy settings:

- Public: The URL for the playlist can be shared; the playlist can be added to a channel section; the playlist will show up in search results and related videos; and the playlist will be posted on your channel page.
- Unlisted: The URL for the playlist can be shared; the playlist can be added to a channel section; the playlist will not show up in search results and related videos; and the playlist will not be posted on your channel page.

- Private: The URL for the playlist cannot be shared; the playlist cannot be added to a channel section; the playlist will not show up in search results and related videos; and the playlist will not be posted on your channel page.

Other settings you can change:

- The ability for other users to embed your playlist on their website or blog. The default setting is to allow embedding. ·
- The ability to have new videos added to the top of the playlist. The default is to place new videos at the bottom of the list.

Subscribing to Channels

If you find a channel/user you wish to subscribe to, click on the "Subscribe" button. Subscribing to a channel will save that link in your subscriptions area. To view your subscriptions, you can click on the link labeled "My Subscriptions" near the top of the left banner. This link will open a page with all of the upload activity from your subscriptions. You can also click on the link to view "All Activity," which will allow you to see any activity that has occurred on videos on the channels to which you have subscribed.

Viewing a Channel Page

If you have subscribed to a channel, you can also view the channel's page by clicking on the title of the channel in the subscriptions area of the left banner. If you have not subscribed to a channel, you can view its page by clicking on the name of the channel (or user) when it appears in your search results or when you are viewing a video. On the channel's page, you will have access to all of the videos posted by that user. Along the top of the channel page you will see links dividing the channel's page into the following subpages:

- Home
- Videos
- Playlists
- Discussions
- About
- Magnifying glass icon (clicking here will allow you to search the channel)

On the Home section of any channel, you will find:

- A suggestion of what to watch next
- Recent activity and recent uploads on the channel
- Top videos for the channel

- Playlists made by the user (Note that these playlists may contain videos from channels other than the channel to which you have subscribed.)

On the "Videos" subpage, you will be able to see all of the videos uploaded by that channel's user. If the box labeled "Uploads" near the top of the page has a down arrow, you can open a drop-down menu to see other video content. This may include videos posted by that user or videos liked by that user. On the right side of the screen, you can change the order in which videos appear.

To view any playlists put together by the user you have subscribed to, go to the "Playlists" subpage. These playlists may be videos from the channel you have subscribed to or from another channel that the user has placed in a playlist.

Next over is the "Discussions" subpage. Clicking on this will allow you to see all of the comments posted on videos from this channel. At the top, there is a place for you to add a comment. And finally, the "About" page will tell you more about the channel creator.

How to Upload Videos to YouTube

To share a video on YouTube, first make sure you are logged in to your account. Then click the "Upload" link on the top, near the right side of your screen, to begin the process. On the next page, you can either click on the arrow icon to select a file from your computer's file directory, or drag and drop the file icon into the box. YouTube accepts the following file types:

- .MOV
- .MPEG4
- .AVI
- .WMV
- .MPEGPS
- .FLV
- 3GPP
- WebM

There is information on the YouTube Help Center about how to convert other file types into one of those listed above.

Once you have selected your video to share, determine the privacy level using the box labeled "public." This will open a drop-down menu with the following choices:

- Public: This means that anyone on YouTube can see your video. It will display in search results, recommended videos, and on your channel.

- Unlisted: This means that only people who you share the link with will be able to view your video. Viewers do not need a Google account, and they can share the link with others. Videos that are unlisted will not appear in search results or on your channel's page.
- Private: In order to share a private video, you will first need to make sure you have a Google+ account. Then you can share private videos with other users you are connected to through Google+. This means your viewers will also need Google+ accounts. Private videos can only be seen by the people you indicate and cannot be shared. They also will not show up in search results or on your channel.

Once you have selected your level of privacy and added your video file, click "Upload." Upload times will vary depending on file size, and you will need to keep this page open until the upload is complete. While the video is uploading, you can give your video a title, write a short description, and if your video is public, list any tags or descriptors you feel would help others find your video. You can also change the privacy setting on the right, add a message to your video, and add it to any of your playlists.

Under "Advanced Settings," you can adjust the following video settings:

- Comments
- License and rights ownership
- Syndication
- Caption certification
- Distribution options
- Age restrictions
- Categories
- Video location
- Recording date
- 3D video
- Video statistics

All of the advanced settings are optional. After the video has uploaded and you have all of your settings in place, click on "Publish." Please note that you can return to edit any of these settings in the Video Information section of your channel.

Editing Video Information

To change or add to your video's settings, visit the Video Manager section. To do this, click on the gear icon at the top. In the drop-down menu click on "Video Manager." Under "Uploads," you can click on the "Edit" link in any of your video's records to edit the settings for that particular video. When you are directed to the editing page for a video, you will see that along the top you can

- edit the video's information and settings;
- add enhancements to the video;
- edit your video's audio track;
- add annotations to your video; and
- add closed captions to your video.

In Video Manager, you can also edit your playlists and liked videos, access any copyright notices from YouTube, and stream live events from your channel. Use the YouTube guide to make sure you have the right equipment and settings selected for live streaming. Below the Video Manager options in the left banner, you are able to access your video's analytics, which can also be accessed through the gear icon at the top of the screen.

CONCLUSION

Using this guide, you now have the ability to use both Google Video and YouTube in your library. Remember that Google Video is great for locating those hard-to-find videos that may not be on YouTube. And YouTube can help your patrons and staff by allowing you to share tutorials and other content, in addition to searching for videos.

REFERENCES

Dankowski, Terra. 2013. "How Libraries Are Using Social Media." *American Libraries* 44 (5): 38–41. http://search.ebscohost.com/login.aspx?direct=true&db=a9h&AN=87453442&site= ehost-live&scope=site .
Purcell, Melissa. 2013. "YouTube and You." *Library Media Connection* 31 (4): 14–16.

Chapter Fourteen

Incorporating Google Trends into Library Administration

Marketing, Outreach, and Collection Development

Amy Handfield and Sarah Cohn

There are two essential areas that contribute to a library's currency: marketing and collection development. In both of these areas, libraries can utilize the data provided by Google Trends. Google Trends can provide an easy, cost-effective method of staying abreast of the cultural zeitgeist for libraries with little or no budget for marketing their services.

TRADITIONAL LIBRARY MARKETING

Looking to other libraries for models of organization and services is a way in which information systems can market themselves. This occurs through networking and discussion with other institutions, such as through conferences and professional organizations. Investigating different library websites and online catalogs is also a useful way to find new ideas; asking users what they need is a viable technique to keep a library up to date and relevant.

GOOGLE TRENDS AND LIBRARY MARKETING

Google Trends and the statistical information that it generates about people's Internet searching patterns is an excellent tool to use when deciding how to adequately serve patrons through social media and library events and programming.

Social media outreach can create connections to how a library's collection and services relate to patron interests and needs. Libraries can use Google Trends Hot Searches and Top Charts to see popular topics searched on the Internet. Subject matter discovered by Hot Searches and Top Charts then can be integrated into a library's social media presence. Hot Searches list the most current popular searches. These search trends can be filtered by time, such as day, week, month, or year, which allows for a high level of accuracy and currency. Google Trends displays how interests change over time, which allows one to determine how appropriate a social media posting is. Top Charts can be searched in Google Trends via a left-hand menu bar that breaks down charts by subject. One can select and browse categories such as "Entertainment and Lifestyle," which provide insight into potential popular interests of patrons.

Google Trends can also aid in determining appropriate language for library promotion and marketing. The Related Searches list shows similar subjects and queries that relate to the term searched in Trends. This is an easy way to expand the number of synonyms that a library uses to promote a program or service. Instead of using library jargon, using terms more likely to be understood by patrons increases the success of engaging library patrons.

For example, if you search "eBook," "Kindle," and "iPad" in Google Trends, you will discover that "iPad" is the most popular search term. "EBook" is a much more library-centric term that is not widely used by the general populace. Considering this, when advertising electronic content that a library has, it is more appropriate to use terms such as "iPad" when explaining how patrons can access electronic information books on their personal devices.

Using Google Trends to examine search tendencies is also a useful tool when developing library programing. This information is useful when deciding if a program or event will be well received by library users. For example, if a library is planning on hosting a storytelling event, they can search the term "storytelling," see the recent sharp interest in the term, and work to create additional storytelling events.

TRADITIONAL COLLECTION DEVELOPMENT

Collection development is essential in maintaining a library's currency and usefulness. A library's holdings must reflect the information needs of their patrons. There are traditional methods of building a collection, such as book reviews and patron suggestions. More recently, acquisition of library materials may come in the form of preselected collections, such as book bundles. Patron driven acquisitions (PDA) is also a current element of some library systems acquisitions protocol for electronic material. PDA systems provide numerous preselected electronic titles to a library's catalog. These items are purchased by the library system only in the event that a patron checks them out.

GOOGLE TRENDS AND COLLECTION DEVELOPMENT

While these typical methods of collection development are considered best practice, they are not necessarily accurate or successful. Google Trends can be used to refine the process. In order to check if a recent purchase adheres to patron needs, the subject matter of the resource can be searched in Trends for popularity and interest. Google Trends can function as a spot-checking device for traditional methods of purchasing material. If a librarian comes across an item in a review or book bundle, he or she can search the title or author to see if it is of popular interest. The popularity of the author can be determined by the interest over time graph. The interactive map that breaks down searches by regional interest is quite useful in deciding if a title is popular in a particular library's location.

In addition to functioning as a secondary tool to improve traditional acquisition models, Google Trends can be used as a stand-alone device for library material selection. The Zeitgeist option is an excellent one-stop shop for popular subjects. Zeitgeist is subdivided into subjects for easy searching and is a convenient way for a librarian to quickly locate popular subjects to build into the collection. There is also an archive of popular searches organized by year going back to 2004, which is useful for researching recent trends in reader or user interest.

Google Trends' Top Charts list the most searched for things, people, or concepts and links to both information about the topic as well as the Google Trends data results. Under "Entertainment," there are top charts for both authors and books that can be used as a direct source for collection development. These obvious choices for top charts provide a snapshot of what and who people are interested in reading. Top charts for other areas such as business and politics, nature and science, and sports can also provide clues for collection development. The Top Charts list for politicians or athletes might suggest new biographies to purchase. The lists for food and travel might suggest new cookbooks or travel guides to add to the collection.

GOOGLE TRENDS AND LIBRARY EQUIPMENT

The use of Google Trends can go beyond selecting monographs and serials. Searching Trends can be useful when comparing different types of media and software that can be utilized in library systems. Using Google Trends to search for software and digital resources is an excellent way to see what is used by others at a given time. In addition to digital resources, Google Trends can function as a way to gauge the most useful hardware and equipment that a library may need in order to function efficiently.

To look quickly at current technology trends, the top chart for tech gadgets is very helpful. This chart provides a ranking of the devices that have been searched. A listed device can be trended by selecting the Explore button located near the thumbnail image of the device. Trends can also be used to compare different types of technology and equipment to forecast what library patrons will use more. For example, if the terms "copier," "printer," and "scanner" are searched together in Google Trends, the user can see that "printer" and "scanner" are searched more than "copier." This search would indicate that an increase in printers and scanners would be a better use of a library's equipment budget than a copier.

CONCLUSION

For libraries interested in providing up-to-the-minute materials and services, Google Trends is a useful tool. Zeitgeist, Top Searches, and Top Charts are just a few functions within Google Trends that provide insight into people's search tendencies. This data can then be put to use in the areas of collection development and library marketing and outreach. Like many Google applications, Trends has changed. Functionalities are being made more robust and new functions are being introduced. As Google Trends continues to expand and evolve, it will remain a valuable resource for librarians.

Chapter Fifteen

Organizing Employees Using Google

Adam Fullerton

Google has a plethora of services that can do almost everything needed in regard to employee management. In the summer of 2013, the Hickman-Johnson-Furrow Library of Morningside College in Sioux City, Iowa, moved its employee communication and organization to the "Googlesphere." We have found it easy to use, quick to access and update, and most of all, exceptional at doing what is needed to manage employee workflows, communication, scheduling, and more.

This chapter will explain how the Hickman-Johnson-Furrow Library set up various Google services to manage its employees. The specific services that will be discussed are Google Sites as an employee portal, Google Calendar for scheduling, Google Drive for document organization and sharing, and Google+ for communication.

WHY CHOOSE GOOGLE?

- It's user friendly while providing numerous avenues for the more experienced user to customize services for specific needs.
- All the services are cloud based, meaning they are stored and accessed via the World Wide Web. Several services offer an offline mode for when you are not connected to the Internet.
- Most services have mobile applications for Android and iOS.

Helpful Tips to Remember

- Make sure to use the same Google account. Changing accounts can lead to sharing and access problems.

- Use the "Help" option located in the settings menu (gear icon) at the top-right corner of all of Google's services.
- If the help option fails, use Google Search to find additional information. Rarely is there an issue that has not already been solved and reported on the web.
- Use Google+ to join various communities dedicated to Google's services. There are hundreds of Google users sharing information and tips and offering free assistance.

CREATE AN EMPLOYEE PORTAL USING GOOGLE SITES

The first service that needs to be discussed is Google Sites. Google Sites, or just Sites, is a versatile website development and management platform. It is important to begin with this service first as it will provide the infrastructure needed to organize the rest of the services and information that will be presented to your employees.

Why Use Google Sites?

- No HTML or CSS required
- WYSIWYG (what you see is what you get) editing
- Easy integration with other Google services
- Private or public settings available

The first thing you will need to do to get started is create a site. After you go through the process of creating a site, several of the most useful and important functions will be discussed. Remember that Sites is a feature-rich service that includes many additional options not mentioned in this brief chapter.

How to Set Up a Google Site

1. Got to https://sites.google.com.
2. Click the red "Create" button at the top left of the page.
3. Several options are presented:

   a. "Select a Template to Use" will allow you to select a predesigned template provided by other Google users. Templates provide a more detailed framework.
   b. "Name Your Site" is for doing just that—naming your site.
   c. "Site Location" is the desired URL for your website.

d. "Select a Theme" allows for the selection of different default colors, layouts, and font styles to be applied to the site. Themes are not as detailed as templates.

e. "More Options" allows you to add a site description for search engine optimization.

4. Click the red "Create" button located at the top left of the page when you are finished.

You are now ready to start editing your new site! To begin, familiarize yourself with the locations of the various tools and menus. There are several different setting menus in Sites. The first to take note of is the gear icon, or "More Actions Menu," located in the top-right corner of the page. Pay particular attention to the following:

- "Page Settings" will allow you to change each individual page's settings for comments, attachments, page descriptions, and several other options.
- "Edit Site Layout" allows you to make changes to the navigation menu, change the header of the site, add a footer to the bottom of the site, and change the default widths of the site.
- "Manage Site" allows you to change the following settings for the entire site:

  - Font style, size, and color
  - Configuring search options
  - Adding analytics to track use
  - Deleting the site from your account
  - Viewing a current list of the existing pages on your site
  - Changing the default theme

- The "Create New Page" menu next to the gear icon is for creating a brand new page for the site. The following information is requested when creating a new page:

  - "Page Name," which will be displayed at the top of the new page
  - "Template," for selecting the purpose of the page, which can vary from storing files, providing lists, or being a standard web page
  - "Select a Location," for adding subpages to already existing pages

- Next to the "Create New Page" menu is the "Edit Page" menu. Click this to open the page editor. Notable menu options include:

- "Insert," for adding images, links, gadgets such as MP3 players, and embedding information from other Google services such as Drive, Calendar, YouTube, and Groups
- "Format," to add headings, quotes, code, and/or alignment of text
- "Layout," for changing the layout of the individual page including columns, introductions, footers, and so forth

Continue to explore Sites as there are many different options and plugins available to expand its effectiveness. Now it's time to start embedding some of Google's other services to make your new portal more effective.

SCHEDULING WITH GOOGLE CALENDAR

One of the major issues our library has to deal with every semester is maintaining our student employees' schedules. In order to keep them organized, we implemented an employee schedule using Google Calendar. Google Calendar continues to prove to be a valuable asset to our staff.

Why Google Calendar?

- Easy to use
- Easy integration with other Google services
- Event searching over large periods of time
- Appointment scheduling
- Mobile apps for Android and iOS

Setting Up a Google Calendar

Calendars can be used for most of a library's internal scheduling. Be creative and try different options to find out what works best for your library.

1. Navigate to https://www.google.com/calendar and log in.
2. On the left side of the page, click the small gray arrow to the right of the heading "My Calendar."
3. Select "Create New Calendar."
4. Enter the appropriate information requested.

   a. Be aware of the sharing settings on this page. Set a calendar to public to make it available to the entire world, or select specific individuals by entering their e-mail addresses.
   b. Be sure to select the appropriate "Permission Settings" for each user.

5. Click "Create Calendar" at the bottom left of the page to finish.

Additional Calendar Functionality

• Change a specific calendar's settings:

  • Click the arrow to the right of a calendar's name under the heading "My Calendars."
  • Click "Calendar Settings" in the drop-down list to change the settings of the calendar.

• To add events and employee shifts to a calendar:

  • Click the red "Create" button at the top-right corner of the page.
  • An event settings page will open.

    • Enter the employee's name for the "Event Title."
    • Set the date and time of the employee's shift.

      • Select the "Repeat" box below the time entry to set up a repeating shift for that employee. This saves time by eliminating the need to enter the same information for different dates in the future.

  • Once all the required information has been entered, click the red "Save" button at the top right of the page.

• To edit an event:

  • Click on the event in the main calendar viewing area.
  • A pop-up message will display, allowing you to copy the event to another calendar, delete the event, or edit the event.
  • Clicking "Edit event" brings up the event settings page.

ADDING A CALENDAR TO SITES

Now that the appropriate calendar has been created for employee scheduling, it can be embedded into Google Sites. If necessary, create a new page on Sites to embed the calendar, or embed the calendar in the "Home" page so employees see the schedule when they first visit the portal.

How to Embed a Calendar in Sites

1. Navigate to https://sites.google.com and open the employee portal site.
2. Navigate to the page that will host the embedded calendar.
3. Click the "Edit Page" button at the top-right corner of the page.
4. Place the mouse cursor on the page where the calendar will be embedded.
5. Click "Insert" from the menu in the top-left corner.
6. In the "Insert" menu, select "Calendar" from the list.
7. This will display all of the calendars currently active in your Google account. Select the appropriate calendar.
8. The next screen will present options for customizing your calendar's appearance. You can try different settings and can always change them if they do not meet your needs.
9. Click the red "Save" button at the bottom of the pop-up to save the calendar's settings.
10. Now click the blue "Save" button at the top right of the page to save the page edits that you have just made. If you do not do this, your calendar will not be saved to the site.

ORGANIZE DOCUMENTATION WITH GOOGLE DRIVE

The next piece to moving employee organization online is to move all relevant documents to the cloud. Google Drive is a cloud-based file storage hub that comes with several other pieces of software for document creation (Google Docs), spreadsheet creation (Google Sheets), and slideshow presentations (Google Slides), among several others.

Why Use Google Drive?

- Easy integration with other Google services
- Third-party add-ons to increase functionality
- Fifteen gigabytes of free storage per account
- Sharing and collaboration of documents in real time
- File conversion options for both uploads and downloads
- Optical character recognition to search inside documents, including PDFs
- Android and iOS applications

Google Drive Introduction

Google Drive can store any file that an account has storage space for; however, it does not necessarily mean that the file will be usable in Google Drive without some form of conversion, which may change the format or quality of the information. For example, when uploading a Microsoft Word document, you will be able to view and share the document in Google Drive, but you will not be able to edit it without converting the file to a Google Doc file. This can often change the formatting of the original document.

It is best to create future documents using Google Docs, Sheets, and Presentation to avoid this issue. Google Drive's products also integrate and come with their own set of pros when working inside of Google's ecosystem. For example, if you have a Google Doc embedded in a Google Site and need to update the content of that document, you will only need to update the document in Drive, and the Site document will update automatically. This is not possible using a Microsoft Word document.

Also, it is good practice to keep a backup copy of every file on a separate hard drive, even when using a cloud storage system like Google Drive. If your Google account was ever compromised, you may lose access to the files stored in your account. It is always better to be safe than sorry, so make a backup for insurance.

Getting Started with Google Drive

- Navigate to https://drive.google.com.
- If this is your first time visiting the above URL, you may be required to answer some questions to reach the Google Drive page.
- Once on the appropriate page, familiarize yourself with some of the basic features.

    - To the top left of the screen there are two red buttons.

        - "Create" is for creating a new Google Doc, Presentation, Spreadsheet, or other file type. You can also create a new folder to organize various files, just like on Windows or Mac.
        - The red button with an upward pointing arrow and a line below it is the "Upload" button. Click this to select specific files or folders from your computer to upload to Google Drive.

- To create a new folder for any existing documents that need to be shared with other employees:

    - Click the red "Create" button at the top left of the page.
    - Select "Folder" from the menu that appears.
    - Name the new folder appropriately.

- Now you have a new folder in Drive. To upload files to this folder:

- Click on the new folder you just created.
- Select the red "Upload" button at the top left of the page.
- Select all the documents you would like to upload to your new folder.
- Before the files begin to upload, you may be asked if you would like to convert any Microsoft Office files, such as Microsoft Word, to a Google file. Remember that this will alter the formatting of the document. If it does not work acceptably, you can always delete the file from Drive and upload it again without converting the file.

- One of the major strengths of Google Drive is sharing and collaborating. To share the newly created Google Drive folder and all of the files stored in that folder, place a check mark in the box to the left of the folder.
- At the top left of the page, a gray button depicting a person with a plus symbol should be displayed once the box is checked. Click this button to open the "Sharing Settings."
- You will find a number of options in these settings.

  - At the top is a link to share the document outside of Drive. This will only work if you change the settings of who has access to the folder. Google Drive has a default of only sharing with "Specific people" (private until you choose to share).

    - Under "Who has access," the top option is "Specific people can access." Click the blue "Change" button to the right to change the "Visibility options" from "Specific People" (private) to one of the more public options.
    - To share with a select group of people, simply enter their e-mail addresses in the "Invite people" text box at the bottom of the screen.

      - You can set each individual's access to "Owner," "Can Edit," or "Can View."

    - Once sharing has been determined and new individuals invited, click the blue "Done" button in the bottom-left corner.

How to Embed a Google Doc, Sheet, or Presentation in Sites

1. Locate the document that is to be shared and set the appropriate "Sharing Settings."

   a. The most efficient way to share a document with a large group of people is to:

    a. Open the "Sharing Settings" for the document.

    b. Click the blue "Change . . ." option to the right of "Specific people can access."

    c. Select the option "Anyone with the link" to allow the document to be viewed in Sites.

    d. Click the green "Save" button and then the blue "Done" button.

2. Log in to your Google Site that you wish to embed the document on.
3. Navigate to the appropriate page that will host the document.
4. Click the "Edit Page" button at the top right of the page.
5. Place the mouse's cursor in the location where you would like to locate the document.
6. At the top left of the page, click the "Insert" menu.
7. Select "Drive" and select which kind of file you will be embedding.
8. From the pop-up window, select the correct document and click the blue "Select" button at the bottom left of the window.
9. Adjust the "Google Document Properties" appropriately. You can change these afterward if the settings do not meet your requirements.
10. Click the red "Save" button at the bottom left of the window.
11. Click the blue "Save" button at the top right of the page to save the document to the page.
12. You have now embedded a Google document into your site.

COMMUNICATING USING GOOGLE+

Google+ has been around for a few years now, but many people are unaware of the power and functionality that it can bring to communication. Gmail is a great tool and e-mail client, but it lacks good group communication functionality. Google+ offers a great group discussion space and excellent online communication options.

Why Google+?

- Private or public visibility options
- Link functionality for important documents and websites
- Various search options
- Organizational tools available
- Android and iOS apps available

Set Up a Google+ Community

In order to set up a Google+ community, you must have an active Gmail account. This is a requirement for all Google+ users. Once you are logged in to your Gmail account, complete the following steps to activate Google+ and create a community.

1. To the top-right corner of your Gmail account, locate the "+You" link and click it.
2. Follow the onscreen prompts to set up your Google+ account.
3. Once your account is set up, move the cursor over the "Home" button on the top-left corner of the page. Select the "Communities" option from the drop-down menu.
4. To join an existing community, you can scroll through the list of different communities at the bottom of the screen.
5. To create a new community, click the blue "Create" button at the top-right corner of the page.
6. Select whether or not you would like a public or private community and click next.
7. Enter a name for your new community. This will be displayed at the top of the community's page, and if public, the community will be searchable by this name.
8. Below the text box for your community's name, there is a drop-down menu that allows you to select whether or not you would like your community to be available in search results.

   a. Select "yes" and anyone on Google+ will be able to search and find your community and request an invite.
   b. Select "no" and you will need to personally invite any community member.
   c. Remember that privacy cannot be changed later, so be sure you select the setting that will be best for your community.

9. Click the blue "Create Community" button at the bottom-right corner.
10. Now it's time to edit the community settings.
11. First, you have the option to change the community's name, if you desire, at the top-left corner of the page.
12. Below the community title text box is the "Tagline" text box. A tagline can be a motto about your library or community.
13. Below the "Tagline" text box is a placeholder for a photograph. Click the gray "Pick a photo" button to upload a photo for your community.
14. After uploading a photo, you will be able to decide which discussion categories to include for your community. Below "All Posts" there are

two arrows. One reads "Discussion" and the second reads "Add category."

    a. The purpose of categories is to organize the community's discussions around a central theme. One possible category could be "Acquisitions" or "Reference."

    b. To change the "Discussion" category, click the word *discussion* and change it to a name that better fits the purpose of that category.

    c. To add a category, click the "Add a category" option below "Discussion" and add the new category.

    d. You can reorder the categories in the list by clicking on an arrow and dragging it up or down in the list.

15. After establishing your categories, there is a large text box to add information "About this community."

    a. If the community is public, this is important to help users understand the purpose of the community.

16. Below the "About this community" text box, you may add additional links to your site. These are displayed to community members at the top right of the community.

    a. Examples of possible links would be the new employee Google Site, a PDF of the employee handbook located on Google Drive, and so forth.

17. You will see one final text box, which is for your community's physical location. If your library has a name similar to other libraries, you may want to include your address so others know they have the correct community for your library.

18. There are two final links below the location text box. One of these allows you to change whether or not your community is found in Google+ search results, and the other allows you to delete your community. Deleting a community cannot be undone.

19. Click the blue "Done" button once you have completed the community settings. You will be able to edit these settings later should the need arise.

As soon as you click the blue "Done" button after setting up your community, you will get a pop-up to share your community. It is most efficient if the individuals you want to invite already have their Google+ accounts activated.

In the "To" text box, you can now add the e-mail addresses of the individuals you would like to invite. As people join your community, you will be able to see all of the current members at the bottom of the left column. You can click on the "See All" link above the images if you have a large group. From this interface, you can also delete members from the community.

Google+ has a number of other features for posting information. You can post YouTube videos and the community will be able to watch the video right from the stream. Searching within the community is possible with the search box located under the community's photo on the left side of the page. As a moderator, a number of options are available to you with individual posts. By clicking the arrow located in the top-right corner of a post, you can

- pin a post to the top of the community so as additional posts are added to the stream, it does not flow down with the older posts;
- edit, delete, or link a post;
- embed a post to your Google site; and
- block comments on a post.

Google+ offers a number of additional features not discussed in this chapter. It is highly recommended that you take time to explore the vast possibilities of Google+. Extras like Google+ Photos, Hangouts, Live Streaming to You-Tube, and more are valuable services provided in the Google+ platform.

As with most technology, Google is constantly updating and creating new functions for their products. Once you dive into using Google services to manage your employees' scheduling, organize internal documents, manage and promote group communication, and create a unique portal, the value of Google really begins to shine. As time progresses, you will be able to make small changes to all of these services to mold and fit your library's needs.

Chapter Sixteen

Using Google Drive for Library Assessment

Seth Allen

Libraries in school and academic settings are increasingly expected to demonstrate how they contribute to institutional goals. Consequently, librarians are looking for more and better ways to measure a number of areas that indicate their effectiveness: benchmarks such as collection size and circulation statistics are no longer adequate measures of effectiveness in schools and colleges. Instead, schools are seeking data that indicates how each department assists the institution to achieve its goals and, in particular, how each department contributes to student learning outcomes (Applegate 2013). Among the free and inexpensive tools available to educational professionals, Google Drive (formerly known as Google Docs) stands out for providing a suite of free tools that librarians can employ in assessment. Using Google Drive, librarians can easily gather assessment data that measures library effectiveness at multiple levels (i.e., class, major, or institution) to gauge students' learning, comfort with research, and satisfaction with library services. This chapter will focus on how school and academic librarians can use Google Drive to effectively assess their instructional services.

GETTING STARTED WITH GOOGLE DRIVE

To take advantage of Google Drive, you will first need to set up an account with a Google product, such as Gmail or YouTube. If you have a Google account, no extra work is needed to use the suite of applications. Simply go to http://www.drive.google.com/ and then click "Go to Google Drive." From there you will select "Create an Account" and fill out the necessary data. Once logged in to your account, you can begin by (1) creating "Google native" documents, presentations, and the like by clicking on the red "Create" button or (2) uploading files created in Microsoft Office Suite by clicking on the red button with the "up" arrow. All files will be converted to the Google Drive format. Presently, there are five types of files that can be created in Google Drive, though virtually all file formats can be stored in Google Drive. Here is a list of document types that can be generated in Google Drive with the equivalent product in Microsoft Office Suite in parentheses:

- Document (equivalent to Microsoft Word)
- Presentation (equivalent to PowerPoint)
- Spreadsheet (equivalent to Excel)
- Form (equivalent to form builder in Access)
- Drawing (equivalent to Paint)

With the exception of Google Drawing, the upload function identifies the file format and converts it into a Google Drive equivalent (e.g., PowerPoint presentations, or .pptx files, will be converted into Google Presentation format). Once you upload or create documents in Google Drive, the forms will be saved by date added. By default, all files are set to private. You can, however, publish these Google Drive files to the web so that others can access them. Detailed instructions will be given in the sections below on how to share assessments and results with others.

ADAPTING ASSESSMENT TECHNIQUES TO GOOGLE DRIVE

Google Drive can be a valuable tool for administering assessments and organizing and visualizing data. This chapter is divided into the following sections:

- Satisfaction Surveys
- Collecting Survey Responses
- Analyzing the Data
- Assessing One-Shot Instruction Sessions
- In-Class Formative Assessments

  - Concept Maps
  - Collaborative Bibliographies

- Analyzing Existing Assessment Data with Google Drive (i.e., surveys, knowledge tests)

  - Grading Rubrics

• Focus Groups

SATISFACTION SURVEYS

Satisfaction surveys are the most ubiquitous form of assessment; they require minimal effort to create, distribute, and collect quantifiable data. Google Forms can greatly assist librarians with collecting data on students' affective outcomes. Within Forms, it is possible to create a visually appealing form that accepts responses in Google Spreadsheets. To begin, open Google Drive and press the red "Create" button. Click on the green form icon. From there, you will see a dialog box to name the form and select a template. The form builder will open and a multiple-choice question will appear by default in this blank form. Select "Question Type" from the drop-down menu, and then select the appropriate type of question. If, for example, you wanted to create a "select all that apply" question, you would select the "Checkboxes" options. Likert Scale questions (i.e., strongly agree, agree, neutral, etc.) can be created using the "Scale" option for one question or "Grid" for multiple Likert Scale questions. The "Paragraph" text option should be used for open-ended questions and the "Text" option should be used for simple one-word answers (e.g., "What is your e-mail address?"). The "Help Text" clarifies the question and appears as a caption under the question in the published survey. Press "Done" and the question will be saved to your form. In keeping with good survey practices, make certain that you

> create multiple-choice answers that are exhaustive (the answers cover all possible options with an "N/A" or "None of the Above" option);
> your multiple-choice answers are mutually exclusive (the answers do not overlap);
> avoid double-barreled questions that actually ask multiple questions (e.g., "Do you think the library should discontinue collecting resources in print to purchase e-books instead?"); and
> avoid negative and biased phrasing (e.g., "Do you think that library instruction is a waste of time?" "Do you think cell phones should not be allowed in the study areas?") (Babbie 2012).

COLLECTING SURVEY RESPONSES

Once you have set up each focus-group question and any corresponding comment boxes, look in the navigation bar above the questions for a tab labeled "Choose Response Destination." This is where you will name the spreadsheet that will collect responses to this survey. Select the second option to create a new sheet in an existing spreadsheet. This will allow you to collect all responses in one spreadsheet but on separate sheets. Once you have created the questions and designated a destination for the responses, you can begin to enter responses in the live form. Click on the "View Live Form" icon. A separate window will open with the live form as it will appear to participants. To send the form directly, click "File" and "Send Form" to send it to the desired participants. To embed the form in a web page, click on "File, then "Embed," and copy the HTML code, resizing the document as needed.

ANALYZING THE DATA

Google Spreadsheets has a number of tools to assist you with compiling and analyzing raw data from surveys. With the spreadsheet containing survey responses open, click on "Form" and select "Show a Summary of Responses" in the drop-down menu. Google Spreadsheets automatically displays answers in pie charts or bar graphs and shows open-ended questions as a block with a few spaces between answers. You can download the data as an Excel spreadsheet by clicking on "File," then click on "Download as," and then select "Microsoft Excel (.xlsx)."

TIPS FOR OPTIMIZING GOOGLE DRIVE WITH SURVEYS

Use the section headers to give specific instructions for a group of similar questions and page breaks to break up long surveys by clicking on "Insert" and "Page Break" or "Section Header" in the edit mode of Google Forms.

Create "contingency" questions using the "Go to the page based on answer" box beside the text of each question. You might want to ask students to clarify their answers to certain questions, such as their satisfaction (or lack thereof) with library instruction. In this case, you will need to create the contingent question on a separate page from the other questions (using the page break in the form's edit mode). Once students answer the contingent question, have all the answers for this question redirect to the main survey form.

Use the Formulas tool to perform instant statistical analyses. In the spreadsheet view, highlight the column you wish to analyze, click "Insert" and then "Function." Select the appropriate function from the list or click on "More" to use a wide variety of tools for analysis.

Take a screenshot of the summary of responses to paste in a Word document, PowerPoint, or web page.

ASSESSING ONE-SHOT INSTRUCTION SESSIONS

Google Forms can be a quick and easy way to assess student learning after library instruction. In a matter of minutes, a librarian can create a form and e-mail the link to a class prior to delivering an instruction session. There are two levels of evaluation that can be effectively captured in the one-shot library instruction session using Google Forms: reaction and learning. Reaction measures students' affective state after an instruction session (i.e., confused, satisfied, and to what extent) and learning (Suskie 2013). Other in-depth, long-term evaluation levels cannot be assessed in the one-shot library instruction session. With these levels of evaluation in mind, you can pair learning objectives with valid instruments for measuring these outcomes.

When developing a post-instruction assessment, try to incorporate all three levels of evaluation. A reaction question might involve using a Likert Scale question. To create a scale question, open up the editing screen of the survey and click on "Insert" and then "Scale." Be sure to define both ends of the spectrum. For example, a question asking students to rate the helpfulness of a one-shot instruction session might label the 1 as "Very Unhelpful" and the 5 as "Very Helpful." In the "Help Text," you might want to indicate that a 3 is equivalent to the "Neutral/Not Applicable" selection. To measure the second level (learning), it is advisable to use a multiple-choice option for single-answer questions or a checkbox option for answers with several potential answers. The third level of evaluation is the hardest of the three to measure within a one-shot session. Generally speaking, behavioral shifts are best measured at the end of a course or program of study. Rubrics are often used to quantify specific behaviors and grade students' ability to perform a given task on a scale. For information on setting up a rubric for the purpose of grading, see the following section, "Processing Existing Assessment Data with Google Forms."

USING SCRIPTS TO AUTOMATICALLY GRADE STUDENT ASSESSMENTS

Google Forms is compatible with a script called Flubaroo, which can be used to automatically grade multiple-choice and short-answer questions. A "script" is a dynamic coding language that instructs web browsers to perform a function, such as computing statistical averages or retrieving articles from databases. With the Flubaroo script, you can set up a knowledge test to administer to students at the end of a one-shot session in Google Forms, distribute it to the class, and have the results graded and sent back to students.

To get started, set up the knowledge test in a Google form and select a spreadsheet destination. If you need help doing this, please refer to the "Satisfaction Survey" and "Analyzing the Data" sections of this chapter. When designing the form, make certain that you include fields for students to enter their name and e-mail address. The script will need this information to e-mail students their grades. Once you have created and opened the spreadsheet where the answers will be collected, click on "Tools" and then click "Script Gallery" from the drop-down menu. In the search bar, type "Flubaroo." Once you locate the script, click "Install" and give authorization access in the pop-up box. Now close the dialog box and you will notice on the horizontal task bar above the spreadsheet (where the File, Edit, View, Insert, etc., functions are) that the last option is "Flubaroo." If this is the case, you have successfully installed the script and are ready to set up grading on an assignment. You will need to click on "View Live Form" and take the exam yourself, filling in the correct answers. This first submission will be used to grade all subsequent submissions. Once you have received students' submissions for the form, open up the spreadsheet with the collected responses and click on the Flubaroo tab. Click on "Grade Assignment" in the drop-down menu and begin the first step of assigning points to each question and designating the field for the student's name. Press "Continue," and in the second step, you will designate the answer key by clicking on the first submission listed (this should be your form submission with the correct answers) and press "Continue." You will see a pop-up box labeled "View Grades" once the tests are graded. On the navigation bar of the spreadsheet, click the Flubaroo icon again and click "Email Grades" in the drop-down menu. In the pop-up box, select the field containing the student's e-mail address and, if desired, add feedback in the box and include the answer key. You can generate a histogram of students' grades by clicking on "Flubaroo" and "View Report."

TIPS FOR OPTIMIZING GOOGLE DRIVE FOR ASSESSING THE ONE-SHOT SESSION

Save the responses of the students in multiple classes in separate sheets of the same spreadsheet. To do this, you will need to create a separate form for each class. In the edit mode of the form, click on "File" and "Make a Copy." Be sure to label the form with the class section. In the duplicate form, click on "Responses" in the navigation menu and then click on "Change Response Destination." In the dialog box, click on the option "New sheet in an existing spreadsheet" and select the appropriate spreadsheet. Google Forms will create a new sheet to collect responses, so be certain to open the spreadsheet and rename the newly formed sheet.

Create a pre-test/post-test using Google Docs. Open Google Drive and press the red "Create" button and select "Form." Be sure to name the form "Pre-Test" or something similar. Once you have set up the form, click on the "Responses" button in the horizontal navigation bar and select "Responses." Now select "Change Response Destination." This is where you will name the spreadsheet that will collect your responses—choose something recognizable, such as "Pre-Test/Post-Test" under the "New Spreadsheet" option. Next, click on "File" in the horizontal navigation, then click on "Make a Copy," and in the dialog box that pops up, name the duplicate quiz "Post-Test." A duplicate form will be created with the new file name. You will, however, need to change the survey's title in the edit mode of the duplicated form to indicate that it is a post-test.

IN-CLASS FORMATIVE ASSESSMENTS

Given the recent trends toward using active learning techniques, librarians and other educators are looking for easy ways to engage students in the learning process. Google Drive has a number of tools that, with a little creativity, could be used as formative assessments. Many school districts and colleges have already switched to Gmail as their e-mail provider. In these school systems, students can readily access Google Drive, or this service can be activated by an IT administrator. If not, it is possible to quickly publish documents online and allow students to edit them by accessing the URL (students do not need a Google account to edit Drive documents). Here are two ideas for in-class activities using Google Docs:

Create concept maps using Google Drawing. Concept maps are ideal for teaching students how to broaden and narrow search terms and think of related words and synonyms in their research strategy. Open Google Drive and click on the red "Create" button and then "Drawing." This opens a blank canvas. You can add shapes and lines by clicking on "Insert" in the toolbar and choosing a circle or line under "Shapes" as needed. To publish, click on "File" and "Share." Click on the "Change" option, and under "Who has access," change the status from "Private" to either "Public on the web" or "Anyone with the link." You can send this link to students before class so that they can edit the document in class. Alternately, you could shorten the link to the drawing using a link truncator service, such as tinyURL (http://tinyurl.com/), and write the shortened link on the board. You can create duplicates of Google Drawings if you would like to organize a class into groups to work on their own concept map. With the concept map open, click on "File" and "Make a Copy" to duplicate your concept map template. Any edits that students make are automatically saved and thus can be graded after class.

"Crowdsourcing" the annotated bibliography. Use the strategy above to create and share a Google Drive document. You could create a document to walk students through the process of finding articles, citing their sources, and briefly summarizing what they found in an annotated bibliography. You could write several hypothetical research assignments on a single document. Each group would be assigned a single research question. After instructing them on how to find articles on a particular database and how to use the "cite" function that many databases offer, you could have them locate an article, cite it in an appropriate style, and write a brief summary from skimming the article. This single document could be reviewed at the end of class for the benefit of the other students.

The options for using Google Drive to conduct formative assessments are endless. As you think of new ways to use Google Drive, you might want to create a folder and name it "Instructor's Toolkit." On the Google Drive home page, click the red "Create" button and then click "Folder." This will allow you to keep and reuse Google Drive documents for teaching in the future.

ANALYZING EXISTING ASSESSMENT DATA WITH GOOGLE DRIVE

Grading Rubrics

Google Drive can also be a useful for creating rubrics to grade students' work. Two such examples would be using simple grading rubrics within Google Forms to grade students' works on knowledge and behavioral tests and entering and coding focus group data. To create a rubric for the purpose of grading students' work, open Google Drive and click on "Create" and "Forms." After naming the form and choosing a template, you can begin to create a simple rubric. Change the default "Text" question to "Grid" question and begin entering the objectives in the "Row 1 Label." Under the column labels, you can type in performance indicators, such as "very good," "good," and so on. One drawback of using Google Forms to create a rubric is that all objectives in a form must use the same labels for performance indicators. Given the varied performance indicators in a typical rubric, it is not realistic to assign labels to performance indicators; one indicator may be "very confident" while an indicator to another question might be "Student finds three or more peer-reviewed sources." Thus, it is advisable to create a detailed rubric in a separate Word document and simply use Google Forms to enter your data and perform statistical analyses. In Google Forms, you might want to code performance indicators on a 1–5 scale, keeping a mental note that 1 is the least desirable score (e.g., "The student is not able to find any peer-reviewed articles on the topic.") and 5 is the most desirable (e.g., "The student is able to find more than five peer-reviewed sources on the topic."). In fact, coding your performance indicators as numerical values is helpful to set benchmarks for instruction (i.e., you want the class average after your presentation to be a "3" and make quantitative comparisons of different classes that are presented the same material in a different delivery style). You could, for example, develop a goal in a one-shot instruction session that students, on average, will be able to find at least three peer-reviewed journal articles on a topic, which translate to a 3 on your rubric. You would then enter the numerical value for each student and quickly average the class. This data could be used to measure the effectiveness of various instruction-delivery styles (i.e., online, in-class lecture, "flipped") for teaching the same content. For an example of how to grade a rubric by coding parameters, see table 16.1.

Once you have graded assignments using Google Forms, you can easily find class averages. Open the spreadsheet of raw data where your grades are being collected and click on "Insert" and then "Function" in the drop-down menu. You will be presented with a list of options for statistical analysis comparable to Excel.

Tips for Optimizing Google Drive for Processing Existing Data

Adapt existing rubrics rather than create one from scratch. Two good websites for this are Rubric Assessment for Information Literacy (RAILS) and Rubistar. The addresses for these sites are http://railsontrack.info/rubrics.aspx and http://rubistar.4teachers.org/, respectively.

Analyze coded data with data analysis software such as SPSS. Download the data as an Excel spreadsheet by clicking on "File," "Download as," and "Microsoft Excel." Spreadsheets can be uploaded directly to SPSS and analyzed in a number of ways.

Install Google Fusion Tables to create custom charts and graphs. To install Google Fusion Tables, open up your Google Drive, click on the red "Create" button, and then click "Connect more apps." Type in "Fusion" in the app search box and install the app. Click "Create" again in the Google Drive menu and now select "Fusion Table." You can create or open existing spreadsheets and create customized charts and graphs comparing data from library surveys and knowledge tests (e.g., self-reported GPA and number of hours per week spent studying in the library).

Table 16.1.   Grading Rubrics Using Google Forms (columns 1–5 contain scores to be entered into the Google Forms rubric)

| Learning Objective | Numerical Value to Put in Google Form | | | | |
	5	4	3	2	1
Student will be able to find ample peer-reviewed resources on any given topic using library database.	Student can find five or more peer-reviewed articles on a given topic.	Student can find at least four peer-reviewed articles on a given topic.	Student can find at least three peer-reviewed articles on a given topic.	Student can find at least two peer-reviewed articles on a given topic.	Student cannot find any peer-reviewed articles on a given topic.
Student will be able to cite their sources in an appropriate citation style.	Student makes less than three errors in citation style, cites all sources mentioned in the body of the paper.	Student makes three or more errors in citation style, but overall has a good grasp of citation practices.	Student makes several consistent errors, indicating some deficiencies of knowledge of citation rules. Student does not cite all sources mentioned in the body of the paper.	Student lists the author and title of the work and URL of websites, but does not use proper formatting for citations.	Student cites no sources at the end of the paper.

FOCUS GROUPS

Focus groups can be tremendously helpful in assessing outcomes in the affective domain (i.e., feelings, comfort level, opinion) and how well students evaluate information (Suskie 2013). Google Forms can be a useful tool for transcribing responses from an audio recording into a spreadsheet.

For basic instructions on creating and saving questions, see the instructions for "Satisfaction Surveys" and "Collecting Survey Responses" above. Since focus-group questions are essentially open ended, you will need to set up the form for data entry with ample space to record responses. Once you have opened Google Forms, click on "Add Item" and select "Paragraph Text" to create a question with a paragraph (as opposed to a single-line box) for capturing responses. You will want to repeat this process for each focus-group question. You might want to include a box in the form to gather notes about the respondents' body language and facial expressions as they answer each focus-group question. To do this, create a "Paragraph Text" question under each question and give it an intuitive label in the "Question Title" box, such as "Comments about Previous Question."

You can either create a new spreadsheet or create a new tab in an existing spreadsheet. If entered correctly, each respondents' answers to a particular question will appear in separate cells in one column. The spreadsheet will have gaps in each column as all subsequent questions will be recorded in the next column, one row below the last response of the previous question. You can copy and paste a cluster of cell data at the very top of each column for ease of analysis. To download this file in Excel, click on "File," then "Download As," and then "Microsoft Excel (.xlsx)."

SUMMARY

Google Drive is a powerful set of free tools that can help librarians perform a range of assessment activities efficiently. Google Forms, in particular, can be used to collect survey data, administer knowledge tests, engage students with hands-on learning activities, and analyze existing assessment data. In a matter of clicks, this information can be shared with library colleagues and stakeholders outside the library. With little effort and creativity, librarians in academic settings can adapt Google Drive to perform varied assessments, collect rich data, and ultimately confirm their value to the parent institution.

REFERENCES

Applegate, Rachel. 2013. *Practical Evaluation Techniques for Librarians*. Santa Barbara, CA: ABC-CLIO.

Babbie, Earl. 2012. *The Practice of Social Research*. 9th ed. Belmont, CA: Wadsworth.

Suskie, Linda. 2013. *Assessing Student Learning: A Common Sense Guide*. Unabridged ed. San Francisco: Jossey-Bass.

Chapter Seventeen

Using Google Forms and Google Drive for Library Use Surveys

Steven Richardson

Of the many applications under the Google umbrella, Google Drive, formerly Google Documents, and Google Forms are among the most underappreciated. Our library found ways to apply these applications to meet our needs in collecting data about the library's usage and the services and materials our patrons wanted. We adopted Google Forms and Drive in both our public services and technical services departments and created usage surveys to determine our patrons' views of our library. Although this might not be ideal for everyone, it can be a great tool within certain limited circumstances.

When we opened our new law school, we found it necessary to begin tracking statistics to demonstrate that the library is an "active and responsive" part of the law school in accordance with ABA standard 601(a). As a small, private university, we lacked some of the funding that would normally be used for purchasing software for tracking library statistics. Furthermore, we couldn't justify spending money on tracking software until we could demonstrate a need, and we couldn't demonstrate a need without some way to track our statistics. To save ourselves from this catch-22, we employed Google Drive to create a survey that we could circulate internally and use to track various statistics.

We began with a survey that would track questions for the reference desk because we required granular data to evaluate the number and types of questions being asked. Using a Google Drive survey, we could input all of that data and transfer it instantly to a spreadsheet. Although some libraries have a spreadsheet with all of that information entered manually at the reference desk, we use laptops and provide mobile reference assistance. As such, we needed a way to share that spreadsheet and have it accessible no matter where in the library we were. Our method, using Good Drive and Forms, proved to be efficient and valuable.

PREPARING THE SPREADSHEET

The circulation desk acts as an information desk for the library. Among other things, the circulation staff explains hours and library policies and gives directions to those who are unfamiliar with our facility. We believed that tracking those questions would provide a more complete view of usage patterns. We could analyze, for example, who was asking for information and what type of information those individuals wanted. With our circulation desk being staffed by student workers who didn't have access to shared computer drives and who had to log in separately to use the terminal (as is so often the case in academic libraries), we also needed a way to enable everyone to input data into the spreadsheet. Using an online survey was the perfect solution, and using a free program meant we didn't have to justify the cost of the software.

Once we began to use the surveys, the benefits of generating more detailed data became obvious. We adjusted hours to better accommodate the needs of the students, focusing our staff time on peak usage hours. This also made our small library staff operate more efficiently, thus enabling us to successfully complete all tasks without requiring overtime or additional assistance. Indeed, when the main campus library, which was using paper to track their reference desk, saw how we were able to make better staffing decisions based on our data collection, they became interested in using Google Drive, too.

INTERPRETING THE DATA

Data collection is not valuable unless it can be interpreted in ways that lead to improved policies. Raw numbers without context are basically useless to anyone but the person who compiled them. In this instance, Google rescued us once again. By looking at the data summaries automatically provided by the survey in Google Drive, the dean for library affairs was able to go into meetings with objective data that everyone easily understood. The graphical displays of the data were easy to interpret and required no additional staff time to generate. Additionally, by showing a constant and steady usage of both desks, the library had detailed information demonstrating its function and value in the law school.

TECHNICAL ISSUES

After the first semester, resetting the form was easy and allowed us to compare the first and second set of statistics. We discovered, for example, that the students were asking more sophisticated questions, illustrating that they had learned to use the law library for more in-depth research and that they had come to use the reference desk for assistance in research.

At the end of the school year, compiling the data and tabulating final results for all of the data into one complete spreadsheet was a simple matter of copying, pasting, and then adding the results from the two end-of-semester summaries. That saved a great deal of time and expense.

We could not, however, use Google for our end-of-the-year report because it would only create a summary for freshly entered, not existing, data. Also, we elected not to use Google for long-term storage of data because there are limits on the number of cells that may be filled in the spreadsheets for a free account at any given time. Furthermore, we wanted data stored on our university servers and backed up on our computers.

EXPANDING TO OTHER DEPARTMENTS

Seeing that the circulation and reference desks had created a simple, efficient, and free way of tracking statistics, our technical services department decided to use the same program. They created a survey that allowed them to receive requests from the faculty. Given that we were mandated to find "green" and paperless alternatives for university offices, expanding the use of Google Drive surveys was an ideal and cost-effective solution.

Our technical services librarian created a survey form based on my work for the reference desk and provided it to the faculty. The survey allowed faculty members to request items related to the courses they were teaching or expedited processing and cataloging of already-ordered items; it also gave them the ability to provide information concerning donated items from faculty or the community. The survey provided an easy spreadsheet that could be exported into the spreadsheets already in use to track that information.

This approach created a direct line of communication between technical services and the law faculty that would not have existed otherwise.

Indeed, given that the login was shared by the technical and public services departments, both departments easily accessed the other's data. Of course, it would have been possible to create two separate accounts were it necessary to have the data separated and to create specific forms to be shared among users if the need had arisen. The idea was considered, but a single login was chosen so that the faculty liaison librarian could access faculty members' requests and to allow the technical services librarian and cataloging specialist (who occasionally provided coverage on the reference desk) to log questions they received.

RESULTS AND ADDITIONAL USES

The results have revealed that the form is used more for logging donations than requesting materials, but the faculty know that they can simply submit their request and have it brought to the attention of the Collection Development Committee.

The form has been valuable for faculty members. It has been used, for example, to request a rush on a particular book needed for a class. Although the same effect could have been achieved by e-mail, doing so would require us to retain all e-mails received and, as we did before, to manually enter data into the spreadsheet. Automating the system thus avoided inbox clutter and increased efficiency.

We also decided to extend the use of our survey forms. In addition to collecting usage statistics, we began to survey our patrons. We constructed student and staff surveys, and the various options for question types proved invaluable in asking questions about satisfaction with our various services, materials, databases, and policies. Ultimately, getting feedback from our students and faculty gave us invaluable information upon which we could improve our services. Of course, surveys of that type are largely based on individual perceptions—and thus somewhat subjective—but it was instructive to correlate perceived versus actual library usage.

The convenience of already having a familiar and integrated program saved hours of time. We didn't have to set up accounts with specialized online survey services, nor did we have to learn new software. We could use the forms for surveys, get the data automatically analyzed, and store it along with the data we already had in the system. Thus, the program required no duplication of effort or additional cost. If we had already had accounts with an online survey system, or had needed to train ourselves on the Google program, it might have altered our calculations.

OTHER POTENTIAL USES

Beyond our current usage, there are a few additional Google products that we have not yet implemented. If we decide to expand our social media presence, we now have a built-in Google+ and YouTube account. If we were not already using Microsoft Office, we could also make use of the calendar and e-mail aspects of the account. If we did not have a website provided by the university, Google Sites would have allowed us to create one. Moreover, although some libraries have a wiki to explain library procedures to student workers or new hires, Google provides that service as well. Considering its flexibility, we will soon determine how to add additional features into our current workflow.

LIBRARIES THAT CAN BENEFIT MOST FROM GOOGLE

While use of Google Drive has benefited our library, not every library will reap the same rewards. A few basic characteristics make a library a good candidate for implementing the use of Google Drive for its basic operations.

Libraries considering the use of Google Drive should consider the following: data and metrics; size, policies, and budget; and access and security. Although Google Drive is convenient, it is not a panacea, and will not meet the needs of all libraries. Other potential software solutions may be more valuable based on each library's consideration of the aforementioned factors.

The threshold consideration is the type of data being collected. Circulation software already produces its own reports, and most libraries are so busy that using Google Drive to track users who enter the building may not be feasible. Reference desk questions, however, are ideal for tracking, and as indicated, using Google Drive provides a better way to collect data, as opposed to keeping hash marks on a page or filling bubbles to count the total number of questions. Ideally you will use a form to track the type of information that can be inputted at your own convenience, and not during a patron interaction.

The next issue is size. A small library is a far better fit for this type of innovation. A single branch, for example, might do well with Google Drive surveys, but a midsize or larger library system will probably collect so much data that a paid Google Drive account will be required. When the price issue arises, it may be worth deciding if dedicated library software would be better suited to your needs. Smaller libraries that may not have the budget for specialized tracking software, or newer libraries who have not budgeted for it yet, might find Google Drive to be an ideal solution.

Older, more established libraries may also encounter issues concerning data collection policies. For ease of use, a library may want to collect data in the same way that has been done in the past so that comparisons may be drawn more easily. Additionally, the benefits of collecting more and potentially useful data must be weighed against the need to establish new baselines and the inability to draw conclusions based on different gathering methods. Other issues relating to particular grants or outside requirements might arise that counsel in favor of a particular method of data collection.

After establishing whether Google Drive is suitable for your library, whether it will fit into your budget, and whether it will meet all of your needs, some consideration should be given to the number of individuals who will have access to the account(s). Questions to consider include: If you are a library system, will each branch track data separately or will everyone funnel information into a single, system-wide account? Do particular departments, such as technical services or reference, have their own accounts or do they join with everyone else?

Beyond the number of accounts and the department or departments to whom they belong, institutions must select the individuals to whom login information will be provided. Relevant questions to consider include: Will the entire department be able to access and change the forms or will that be restricted to a particular person? Will the IT department set up everything and send out reports on the data, or will each department do that on their own? If too many people have access, how secure is your data? Given that this is cloud storage, is that sufficiently secure for your needs? (The issue of security of data is especially important for some special libraries, particularly those in the legal, medical, and business fields.)

Once a library has carefully considered these issues, it can make an informed decision regarding the value of implementing Google Drive. At our institution, our small size and semi-autonomous nature meant that Google Drive was ideal. We were able to create the data collection policies without worrying about the policies of other libraries or departments. Because we did not already own software to track our statistics and had not included a budget request for any in our yearly planning, we had the freedom to choose the best solution for our needs.

Moreover, as a new school, we did not have to concern ourselves with past data and how to reconcile it with a new system. We were also able to select the individuals who would have access to the account and gave each of those people plenary power to create and modify their own forms as they saw fit. Of course, this does not mean that Google Drive can only be used by a library like ours. Libraries of different sizes and with different needs can also benefit from Google after considering the above factors.

ADDITIONAL CONSIDERATIONS

If a library decides to implement Google Drive, there are a few technical issues that warrant consideration. The least important matter is the technology. Virtually all libraries nowadays have computers with Internet access. With Internet access, setting up a Google account simply requires a name and a phone number. I would recommend that you make sure that your name in some way correlates to your library, so that you can, if you so choose, use the account for more public functions.

The flexibility of Google Forms immeasurably enhances data collection. We made extensive use of the drop-down responses because it provided clear and distinct categories that would produce valuable and relevant statistics. Not all questions, however, can be tracked in this manner. To keep a record of the types of questions being asked, and the questions themselves, we included two paragraph text boxes, one for the patron's question, and one for our response. In libraries where there are more significant concerns about privacy, storing data, especially data relating to patrons, in the cloud might be perceived as a risk; thus, limiting responses to predefined categories might be a better option.

In creating the usage surveys, we decided to replace the drop-down options with a multiple-choice answer. We chose to do that to make it much easier on the user to find the answer they wanted. The upside was greater ease of use for the respondents. It also meant that the results displayed in a bar graph based on the numbers of responses, and not the percentages. We also added short-answer questions to allow people a place to explain their answers if they felt that the preselected responses were insufficient or incomplete.

Libraries must also determine the type or types of data to track. Knowing the time and date of a question, for example, can help determine peak usage times for staffing purposes. Tracking how patrons interact, whether by phone, e-mail, instant message, or in person, can also help identify the library's technical needs.

Some people, though, can be tempted to go overboard and try and track everything about a patron: age, gender, primary language, and so on. Knowing the purpose of each category of data, therefore, is a threshold consideration. Indeed, if you are not planning on adding staff with particular language skills, knowing the primary language of a patron is useless, and creating an unnecessarily long form simply to collect additional data wastes everyone's time.

In addition, testing the various types of questions before sending a form out to users, or even circulating it internally, ensures that they evoke the information sought. The questions should elicit useful responses; testing beforehand allows for modification and revision as necessary. Librarians always have the option of clearing responses or simply creating a test form where you can revise options without fear of making a mistake.

Also, being aware of length and complexity guards against an overly long summary. If you decide to add patron questions or responses it may generate long results very quickly, since they will be reproduced, in full, in the summary sheet.

When you do decide to take your form live, you will want to consider how many people you want to send it to and how you will send it to them. In several cases, we found it useful to e-mail the form to ourselves and then forward that e-mail to our patrons. This gave it a more official appearance because it came from an e-mail that was already known to be associated with our library, whereas our Gmail would have been a totally unknown address and potentially caught in spam filters. If you want to send the form to an entire group in your e-mail, the same solution might be necessary. If, however, you only want to send the form to a small number of people, and they are all expecting the e-mail, sending it directly might suffice.

Finally, remember that the forms are not permanent. You can add or delete questions as appropriate. If you find that a response category is no longer apt, it can be removed. If you decide later that you want to change a question, you can do so. It might alter your responses, but it is always an option. Likewise, if you decide that you do not want to delete a form or survey but do not want people answering it for a while, you can simply stop accepting responses. As your data collection needs change, you can, and should, update the form to reflect that.

CONCLUSION

Our experience has demonstrated that, if used correctly, Google Forms and Google Drive can be very valuable to a library that wants to collect good, quality data. Although they are not the right solution for everyone, Google Forms and Google Drive are sufficiently flexible that many libraries looking to reduce or eliminate paper forms might find them very useful. Despite the various issues that must be considered, all librarians should at least consider whether Google Drive and Google Forms can be valuable additions to their institution.

Part IV

Collection Management

Chapter Eighteen

Google Books API

Adding Value to Print Collections

Andrew Weiss

INTRODUCTION: ACCESSING THE VALUE OF THE PUBLIC DOMAIN

One of the most exciting aspects of the Google Books digitization project is the en masse scanning and dissemination of public domain works. Though exact figures are elusive, Google Books currently contains approximately thirty million books with at least one million in the public domain. In comparison, the HathiTrust, a massive digital library based partly on Google Books' digitized corpus, contains 3.7 million public domain volumes. The Internet Archive contains nearly a million works in its collections, all of which are in the public domain. This scale of access to open cultural artifacts is unprecedented. It follows that emphasizing these openly accessible public domain books should be important to all libraries. It would be of great use for a library, then, to supply links to these public domain books in records found in their online catalogs to patrons who might not know that they exist. Indeed, the greater the accessibility to public domain books, the better patrons' information needs can be met. Most library users in the digital era are now accustomed to being able to access works immediately and without barriers. Such demands have become commonplace and provide greater satisfaction to users. As a result, providing a virtual copy to browse may instigate greater trust in the user. It might then result in greater frequency of accessing library resources in the future.

ONLINE CATALOG INTEGRATION VIA GOOGLE BOOKS API FAMILY

In a real-world example, libraries can use Google Books to link their integrated library system (ILS) with online, openly accessible public domain texts by adopting the Google Books API Family. Briefly, an API (application programming interface) in computer programming delineates a set of functions or routines to carry out specified tasks. The tasks are human readable and determine how certain software components interact with each other. The Google Books API Family, specifically, is a web-based API, which means that it is a set of HTTP request messages built upon a representational state transfer (REST) web resource. The Google Books API Family functions, then, as a suite of various programs that will allow one to embed specific Google Book titles within a web page. The API also allows users to create specialized book searches and establish query parameters, retrieve specific volumes from Google Books, retrieve a list of books on a public Google "My Bookshelf," and add or remove volumes from "My Bookshelf."

In the example used at California State University, Northridge, Oviatt Library, the API is embedded in the display record of the online catalog. A record, shown in figure 18.1, in the library's Millennium ILS shows the item record for *Notes on Novelists* by author Henry James (http://suncat.csun.edu/record=b1162615). Written in 1914, the book is in the public domain and thus fully visible in Google Books. This is indicated on the screen by a "full text" button linking out to the record in Google Books. This enhanced ILS

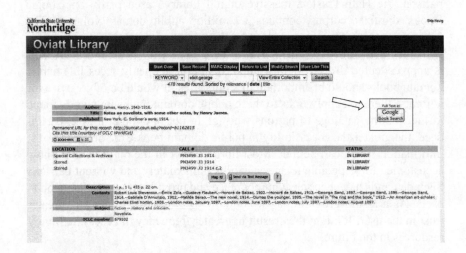

Figure 18.1. Screenshot showing CSUN Oviatt Library Millennium ILS with an embedded Google Books "full text" link button.

record allows students to directly read a book that the library may have without having to examine its contents in the stacks.

Sometimes, if a book is checked out, the student might be able to access this book in Google Books and read it this way. This feature is yet another way to help students access a book prior to coming to the library as well. For organizations with numerous online users, such as a university with an online college or degree program, remote access will help those students who live at a distance from the physical location.

The JavaScript coding for the Google API is relatively simple. Upon receiving a valid response from Google, JavaScript generates a URL for the book record within Google Books. It then inserts the URL into a <div> section in the Millennium ILS's HTML page coding as follows:

```
&lt;div id=google style="font-size:9px;margin-left:4px;"&gt;
&lt;span class="book-title" align="center"&gt;&lt;/span&gt;
&lt;/div&gt;
```

The JavaScript queries the Google Books database for a matching book title. If the title queried has a full text to provide, the cover of the book as well as a button linking to Google Books appears in the online catalog.

OBSERVATIONS FROM USAGE

Embedding the API into the catalog has proven to be a useful tool, especially during reference interviews. Reference librarians go into Google Books to get a sense of the types of books or content that the library patron would be interested in accessing. This has turned out to be a time saver for many patrons as well as reference librarians. Even just seeing the thumbnail of the book cover has helped patrons confirm the identity of volumes they were seeking.

Some issues have been observed, however. One issue has been an uneven user experience. Not all library books are in the public domain or available in full view. As a result, users are sometimes provided a link to the book, but many times they are not. This inconsistency confuses users. The changes in access are also subject to Google's whims. Perpetual access is not guaranteed, and outside forces can sever the links to the resources. The inconsistency includes books that are technically in the public domain but have not yet been digitized.

Another issue is the varying levels of access in Google Books itself. Unlike the Internet Archive, which allows access to everything in its collection, or the HathiTrust, which provides two levels of access, Google Books provides four levels of increasing access. The first level is merely an item record no different than an ILS record, except for the less robust metadata.

The second level is called a "snippet view." This view allows for readers to see several small excerpts from the book, usually amounting to no more than a sentence or two. The third level is called a "preview." This view allows readers to see a limited number of full pages at a single viewing. Usually it's limited to the covers, tables of contents, and the first several pages or so. Sometimes a random search of preview pages is allowed as well. Finally, the fourth level is a full access view of the material. Public domain materials fall into this, and most of the items appearing in the API box in the ILS are culled from these materials.

CONCLUSION

The Google Books API Family can provide a lynchpin for many libraries as they attempt to generate interest in their physical library collections as well as begin to develop strategies for providing access to the e-book. With the development of linked data, library catalogs and records are being dissected to help form a wider, more accessible world. Integrating and championing the public domain will be an important part of the added value generated by the mass aggregation, mass linking, and mass scanning of our legacy print monographs.

REFERENCES

Google Books. 2014a. "Google Books API Family." https://developers.google.com/books/? csw=1.
———. 2014b. "Google Books Library Project." http://books.google.com/googlebooks/ library/.

Chapter Nineteen

Google Books as a Library Resource

Amanda Dinscore

Since 2004, Google has been digitizing the full text of books from libraries all over the world. Despite lawsuits, criticism, and predictions that the project would mean the end of libraries as we know them, Google Books (previously Google Books Search and Google Print) has been a boon to researchers of all levels of search expertise.

A long-running lawsuit filed against Google by the Authors Guild in 2005 brought a great deal of attention to the project. The eventual dismissal of the case by Judge Denny Chin on November 14, 2013, generally was considered a huge victory for both libraries and scholars. It allows Google to continue supporting the discovery of books that are both in and out of copyright and, for some titles, to make short sections, or "snippets," of the text available to the public. Despite the Authors Guild's claims that this is a violation of copyright, Judge Chin agreed that, rather than allow users to pirate copyrighted material, Google Books is a valuable tool that helps users identify and find books, easily determine their usefulness, and locate the nearest library or bookstore where they can check out or purchase the book in order to read the full text if the title in question is still under copyright (Chant 2013).

Google states that its mission is to "organize the world's information and make it universally accessible and useful" (Google 2014a). Throughout history, books have served as the primary way in which knowledge is transmitted but have often been inaccessible to a wide audience of readers who did not have access to a library or who could not afford to purchase books themselves. Google Books, as well as book digitization projects at the Internet Archive and elsewhere, provide access to the content of entire libraries to anyone with an Internet connection. The potential for individual discovery, connecting ideas, and innovative uses of digitized texts is just beginning to be realized. Partner libraries at institutions such as Cornell, Harvard, Oxford, Stanford, and many other academic libraries around the world have made Google Books one of the most ambitious projects to increase access to the world's knowledge that has ever been embarked upon (Google Books 2014a). While many books still under copyright require purchase or check out from a library to view the full text, Google Books allows for increased discovery of texts that might otherwise go overlooked and undiscovered.

CONCERNS AND CHALLENGES

There are, however, concerns that allowing a for-profit company to monopolize responsibility for digitizing and storing the world's book content is risky. Nonprofit organizations such as the Internet Archive have also embarked on large-scale digitization projects. The Internet Archive now makes over two million scanned books available online and has partnered with many of the same high-profile institutions that Google has worked with—including the New York Public Library, Harvard, the University of California, and others—to scan its own collection of books ("Digitizing Print Collections" 2014). However, the high cost of digitization and the necessary access to such a large number of books has prohibited most libraries and organizations from being able to scan books in such large numbers. Libraries that have partnered with Google to scan their holdings are provided with a digital copy of the book for their own patrons' access and institutional archive. Furthermore, many of these partner libraries have collaborated to form the Hathi-Trust Digital Library, an online repository of content scanned by Google, the Internet Archive, and by the libraries and archives themselves. There are currently over eighty institutions participating in the HathiTrust project, and it hosts over 5.6 million book titles. Membership, open to institutions worldwide, provides patrons with access to content that may not otherwise be available in Google Books (Eichenlaub 2013). Furthermore, the recent launch of the Digital Public Library of America (DPLA) in spring 2013 is an attempt to unify access to digitized collections and make them accessible to everyone, free of charge, via a single portal (Darnton 2013). Undoubtedly, the mass digitization of the world's book content is a goal that many organizations are helping to make a reality. And, while Google Books is at an advantage due to its size and financial assets, others are also making significant headway.

Privacy is another issue that users of Google Books may want to learn more about. Google states that in an attempt to enforce copyright laws by limiting access, it does keep track of users' page views and, if you are signed in to your Google account while using Google Books, it will associate these page views with your personal information (Google 2014b). Individuals concerned with privacy while using Google Books Search and other Google tools should review the company's privacy policy, available at http://www.google.com/policies/privacy/.

One particular challenge to utilizing materials scanned by the Google Books project is the quality of the scans and metadata assigned to the books. The project uses optical character recognition (OCR) to convert the scanned images of the books' text so that it is machine readable (Eichenlaub 2013). As one would expect, this can make the process of searching the full text of books problematic. Human error can also sometimes lead to distorted images, scanners' hands captured in the image, and other peculiarities, many of which have been documented in an online blog titled *The Art of Google Books*. A 2012 study analyzed a sample of four hundred Google Books records and revealed that 36 percent contained metadata errors. Incorrect publication dates, misspelled authors' names, and links to the wrong books can make locating a specific text very frustrating for users (James and Weiss 2012). Despite poorly scanned images and inaccurate metadata, however, the benefits of Google Books to scholarly research certainly outweigh its limitations.

GOOGLE BOOKS CONTENT

Current reports state that Google has scanned more than thirty million books from over forty academic libraries around the world (Darnton 2013). While many may be duplicates or different editions of the same title, that still leaves an incredible number of books to be discovered by researchers. Since these books primarily come from academic libraries, you probably won't find your latest *New York Times* fiction bestsellers, but you will find a lot of books that are excellent for scholarly research on just about any topic.

Additionally, many library partners have allowed Google to digitize archival materials held in special collections, materials that previously could only be viewed in person. Google Books has partnered with libraries around the world to digitize a great deal of primary-source material in a variety of languages, including many texts that were previously inaccessible to scholars. In addition to this wealth of primary-source material, you can also search the full text of hundreds of magazines. Many of these magazines only provide full-text access to a limited number of issues, but one notable exception, *Life* magazine, offers full access to over 1,860 issues from the 1930s to the 1970s.

COPYRIGHT AND FULL-TEXT ACCESS

Before you begin your search, it's important to know in advance what limitations exist for accessing the full text of books you find in Google Books. Books that have been digitized through the Library Partner Project may have expired copyright. This generally applies to books that were published prior to 1923 and are now considered to be in the "public domain." In this case, the full text of the book is usually available. You can view the entire book online or even download it in PDF or EPUB format to read offline. H. G. Wells's *The Time Machine*, L. Frank Baum's *The Wonderful Wizard of Oz*, and Alexander Dumas's *The Three Musketeers* are just a few of the many wonderful books that are in the public domain.

If books are under copyright, however, and depending on agreements with authors and publishers in the Google Books Partner Program, only a "snippet" view of the full text may be available or, in some cases, only basic information about the book is included with your search terms in the context of a few sentences of text. In order for Google to monitor copyright and limit page views, you may have to log in with your Google account to obtain access to the snippet view (Google 2014b). For these books, Google Books serves as a gateway to locating the full text in print or e-book format from libraries and booksellers.

FINDING THE FULL TEXT

To start searching Google Books, go to the home page at http://books.google.com/. If you know exactly what you are looking for or want to perform a more precise search, you can also start your search from the Advanced Search page at http://books.google.com/advanced_book_search. From Advanced Search, you can limit your results to only books that are available in full-text view, search for magazines only, or limit by date, author, and subject.

If the book you are looking for is in the public domain, you may still find other editions that are not. For example, a search for *Wuthering Heights* by Emily Brontë brings up many editions of the book, most of which are not available in full text. But, by limiting to full view from the Advanced Search page, you can limit your search to just the full-text editions. A button on the top left of the item page will let you know that it is a free e-book and, when you hover over it, will state which devices the text can be read from, formats in which it can be downloaded, and special features of that particular book.

For items that are not in the public domain, you will see in the item record a button that states "Get Print Book." From here, you will see a link to sellers, such as Amazon, where you can purchase the book. You will also see a link that says "Find in a Library." Click on this link and you will be directed to WorldCat, an online catalog of libraries' holdings from around the world. The book's information will be there and, if you scroll down, you'll see the libraries nearest to you (based on your IP address) where you can check out the book. WorldCat also has an excellent citation tool. Click on "Cite/Export" at the top right to have WorldCat generate the citation for your book in a variety of citation styles.

NAVIGATING GOOGLE BOOKS SEARCH RESULTS

Once you find a book, there are several features that are worth exploring. Depending on author and publisher agreements, copyright, and other issues, not all features will appear for every book.

If you do get a snippet or full-text view, you'll be taken into the text of the book, and your search terms will be highlighted within the text. You can also navigate through the book based on where your search terms appear using the yellow bar above the text. You'll also see a search box where you can change your terms and search the full text of that particular book only. This can be useful when you're trying to determine if a particular text will be useful to you in your research and whether you should consider checking it out from the library or purchasing it.

Click on "About This Book" for reviews, selected pages, the table of contents, keywords, related books, popular passages, and bibliographic information. From here, you can also export the book's information to a citation management program such as EndNote.

MY LIBRARY

The My Library feature is an excellent way to keep track of all the books and other content you discover using Google Books. It allows you to create your own "library" where you can create and organize a personalized selection of material and make notes and annotations. You can create and mark specific virtual bookshelves as public or private and share them with others. You need to have a Google account to use My Library and can register for one at http://www.google.com/accounts.

To add books to your library, simply search Google Books and, when you find a book you want to add, click on "Add to My Library" and choose the bookshelf where you want the book's information listed. When you have My Library connected to your Google account, you can also submit book reviews.

PRACTICAL USES FOR GOOGLE BOOKS

Is a book you want at your library checked out or otherwise unavailable? See if you can place the item on hold with your library, but then see if there is a snippet view in Google Books, and start reading it there. Sometimes all you need is a chapter, and, if you're lucky, that entire chapter may be available in Google Books.

Do you have an article or book and you want to know who has cited it? Google Books can be useful in locating book and journal article titles, or even authors' names, within the full text of the books it has scanned. Remember that you are searching bibliographies and in-text citations to the source you have in mind, so format your search accordingly. Use the Advanced Search page and do a phrase search for either the author name or article or book title. You can also use Google Scholar's Cited By feature to expand your search.

Did you try using your library catalog to search for books on your topic and not get many relevant results? Keep in mind that your library's online catalog search box usually only searches the title, bibliographic information, and sometimes the table of contents of books. If you don't limit your search terms to the most essential aspects of your topic, or if your topic is too narrow, you may not end up with many relevant results. Try to think of the broader context in which your topic appears and search for that. If you're still not having any luck, try Google Books. Because Google Books searches the full text of scanned books, rather than just the title and bibliographic information, you will get many more results. Then, if the full text isn't available, click on "Find in a Library" to get back to your library's online catalog. Or, you can simply copy and paste the book's title into your library's catalog search box.

Are you just getting too many results to weed through? This can be an issue with Google Books because, as mentioned previously, you are doing a search of the *full text* of millions of books. To narrow your search results, go to the Advanced Search page and use some of the search fields to limit your search. The Subject field can be particularly useful to narrow your search to a particular subject area or field of study. There's a fine line between narrowing your search to find more relevant results and narrowing it so much that you don't get *any* results, so make changes, try again, and see what yields the best results.

Do you have a quote in mind, but you can't remember the exact wording or don't know where it came from? Use Google Books' Advanced Search page to search for some of the words as a phrase or enter the phrase in quotation marks in the main search box. From there, you can often find books in which the quote was used and, if you're lucky, track down the original quotation and its author.

NOW THAT WE HAVE GOOGLE BOOKS, DO WE STILL NEED LIBRARIES?

Google Books is an incredible resource for the discovery of books and magazines but, as this chapter has shown, it in no way serves to replace libraries. In fact, libraries can do a lot to publicize and open up their own content collections through Google Books and other Google tools. For example, many libraries conduct workshops and create online tutorials on how to use Google tools effectively. By showing users how to use Google Books to connect to the library's own resources, librarians and library staff can help bridge the gap between an online tool that users are already familiar with and library content that may be underused. A conversation on how to access the full text of books online can also provide libraries with the opportunity to introduce users to their own e-book collections.

As Google states, its ultimate goal for Google Books is to "work with publishers and libraries to create a comprehensive, searchable, virtual card catalog of all books in all languages that helps users discover new books and publishers discover new readers" (Google Books 2014b). In this way, Google Books shines a light on many out-of-print, obscure, and underused resources that have been housed in libraries and archives for decades. Through its ability to search the full text of scanned books, it can provide scholars with an entry point into relevant texts, available online or accessed through a library, that may have previously gone undiscovered.

REFERENCES

Chant, Ian. 2013. "Lawsuit against Google Books Dismissed." *Library Journal* 138 (21): 14.

Darnton, Robert. 2013. "The National Public Digital Library Is Launched!" *New York Review of Books*. Last modified April 25, 2014. http://www.nybooks.com/articles/archives/2013/apr/25/national-digital-public-library-launched/.

"Digitizing Print Collections with the Internet Archive." 2014. Accessed April 13, 2014. https://archive.org/scanning.

EBSCO*host*. 2014. *MAS Ultra School Edition*. Accessed March 25, 2014. http://www.ebscohost.com.

Eichenlaub, Naomi. 2013. "Checking In with Google Books, HathiTrust, and the DPLA." *Information Today* 33 (9): 4–9. Accessed March 25, 2014. http://www.infotoday.com/cilmag/nov13/Eichenlaub--Checking-In-With-Google-Books.shtml.

Google. 2014a. "About Google." Accessed March 25, 2014. http://www.google.com/about/.

———. 2014b. "Does Google Keep Track of the Pages I'm Viewing or the Books I Read?" *Google Books Help*. Accessed March 25, 2014. https://support.google.com/books/answer/43733?hl=en&ref_topic=9259.

———. 2014c. "How Do You Determine If a Book Is in the Public Domain and Therefore out of Copyright?" *Google Books Help*. Accessed March 25, 2014. https://support.google.com/books/answer/43737?hl=en&ref_topic=9259.

———. 2014d. "Privacy." *Google Policies and Principles*. Accessed March 25, 2014. http://www.google.com/policies/privacy/.

Google Books. 2014a. "Library Partners." Accessed March 25, 2014. https://www.google.com/googlebooks/library/partners.html.

————. 2014b. "Library Project." Accessed March 25, 2014. http://books.google.com/googlebooks/library/.

James, Ryan, and Andrew Weiss. 2012. "An Assessment of Google Books' Metadata." *Journal of Library Metadata* 12 (1): 15–22. doi:10.1080/19386389.2012.652566.

Chapter Twenty

Google Ngram Viewer

Andrew Weiss

INTRODUCTION: GOOGLE, MASS DIGITIZATION, AND THE EMERGING SCIENCE OF CULTURONOMICS

Google Books famously launched in 2004 with great fanfare and controversy. Two stated original goals for this ambitious digitization project were to replace the generic library card catalog with their new "virtual" one and to scan all the books that have ever been published, which by a Google employee's reckoning is approximately 129 million books (Google Books 2014; Jackson 2010). Now ten years in, the Google Books mass-digitization project as of May 2014 has claimed to have scanned over thirty million books, a little less than 24 percent of their overt goal. Considering the enormity of their vision for creating this new type of massive digital library, their progress has been astounding.

However, controversies have also dominated the headlines. At the time of the Google announcement in 2004, Jean-Noel Jeanneney, head of Bibliothèque nationale de France at the time, famously called out Google's strategy as excessively Anglo-American and dangerously corporate centric (Jeanneney 2007). Entrusting a corporation to the digitization and preservation of a diverse worldwide corpus of books, Jeanneney argues, is a dangerous game, especially when ascribing a bottom-line economic equivalent figure to cultural works of inestimable personal, national, and international value.

Over the ten years since his objections were raised, many of Jeanneney's comments remain prescient and relevant to the problems and issues the project currently faces. Google Books still faces controversies related to copyright, as seen in the current appeal of the *Authors Guild v. Google Books* decision (Authors Guild 2013); poor metadata, as pointed out in several studies, including Geoffrey Nunberg (2009a; 2009b), Ryan James and Andrew Weiss (2012), and others; poor scanning as described by Ryan James in 2010; the destabilization of current markets, as pointed out by the Authors Guild (2013) and others such as Jaron Lanier (quoted in Timberg 2013); and, finally, the general problem of representative diversity in their digitized corpus, as mentioned by Andrew Weiss and Ryan James (2013).

Despite these major obstacles and problems, the Google Books project has continued without cessation even as other well-funded projects such as the Microsoft-Yahoo book digitization project, for example, have disappeared (Arrington 2008). Google Books' longevity is partly a result of the passion of Google's leaders as well as the deep resources that spring from them. The project has even lasted *within* Google. This is no small feat as many well-known and beloved projects and services have been discontinued by the company over the seventeen years it has existed. The shortlist of discontinued services includes iGoogle, Google Answers, Google Wave, and Google Reader.

But Google Books has staying power. A major reason for this longevity and potential long-term influence may have less to do with individual access to particular titles within its collections and more to do with the overall data sets—especially metadata and indexes—generated by the project. These openly available data sets are being used by researchers for a new kind of research named *culturonomics*. This chapter will explain how libraries and librarians can make use of the research procedures established by culturonomics and the tool developed for it, the Google Ngram Viewer, to aid their patron services.

CULTURONOMICS: A BRIEF DESCRIPTION

Jean-Baptiste Michel and colleagues (2011) have been mining the publicly available data sets from Google Books to "track the use of words over time, compare related words and even graph them." Their research involves examining the frequency with which words appear over time. The researchers examined 5,195,769 books, roughly 4 percent of all printed books, and measured the frequency of appearance of terms across variable times (Michel et al. 2011). Some caveats with this method exist, however. As a result of the small sample, variations and fluctuations in the overall corpus may exist. Furthermore, it is often unclear—and Google hasn't fully disclosed this— just how many texts are available online. Although thirty million is offered as a figure by Google, this is not verifiable. Additionally, the Google Books corpus does not include newsprint, maps, or other nonmonographic materials, which account for a sizable part of library collections.

Nevertheless, the print monograph record can surely be used as an aggregate mirror to peek into the culture as a whole, but one must not forget that it is still one slice of a larger mass, some of which—such as physical experience, unrecorded experience, spoken word, so-called deviant and/or censored material, illegal "black market" material—may never be captured to a satisfactory degree in digital formats. Additionally, works subjected to optical character recognition (OCR) software can be rendered illegible if the fonts are not standard (a true problem for scripts of non-Roman letters such as Japanese) or if other conditions such as faded ink, poor paper quality, and marks or blemishes reduce the digital image quality.

Despite some of these limitations, the result of Michel and colleagues' study has been increased interest in digital humanities and "culturonomics," which is defined by the researchers as an approach that relies on "massive, multi-institutional and multi-national consortia[,] . . . novel technologies enabling the assembly of vast datasets . . . and . . . the deployment of sophisticated computational and quantitative methods" (Michel et al. 2011).

THE GOOGLE BOOKS NGRAM VIEWER

Developed out of the Michel et al. experiments, the Google Ngram Viewer represents a new phase in the use of print materials and demonstrates the power of the added value that a digitized corpus can provide to researchers. Users are especially benefited by its search and data-mining capabilities. Historians, writers, artists, social scientists, and librarians will find the tool useful.

The Google Books Ngram Viewer corpus was created by aggregating five hundred billion words from various monograph/book materials (excluding serials) found in the Google Books collection. The Ngram Viewer defines terms and concepts as "grams." A single uninterrupted word, such as *banana* would be considered a 1-gram; *stock market* would be a 2-gram; *The United States of America* would be a 5-gram. Overall, conceptually speaking, a sequence of any number of 1-grams would be called an *n*-gram. According to Michel, books scanned during the Google Books digitization project were selected for their quality of OCR (optical character recognition) and then indexed. The corpus breaks down by language as follows: English: 361 billion; French: 45 billion; Spanish: 45 billion; German: 37 billion; Russian: 35 billion; Chinese: 13 billion; Hebrew: 2 billion. As shown in figure 20.1, by far the largest amount of *n*-grams represented in their sample is English. This suggests that the search would be somewhat compromised in terms of diversity and representation of different cultures and languages (Michel et al. 2011).

USING THE NGRAM VIEWER: JAPANESE AUTHORS AND THE "NOBEL BUMP" IN NOTORIETY

I conducted several searches using the Ngram Viewer (https://books.google.com/ngrams) to demonstrate some of the power, as well as the impact on libraries, the tool can have in the study of current trends in digital librarianship and scholarship, including the digital humanities. In the first example, shown in figure 20.2, a search using the names of three of Japan's most well-known authors, Yasunari Kawabata, Kenzaburo Oe, and Haruki Murakami, was conducted. The search results show the frequency with which the words appeared in the English language, both British and American, during an approximately eighty-year period between 1930 and 2008.

For the first sixty years, Kawabata is the most mentioned author among the three. From 1930 to 1970, Kawabata was the most well-established author; Oe's first publication did not occur until 1957; Murakami did not start writing until 1978. Interestingly, there are large spikes in the mention of these authors corresponding to their association with the Nobel Prize for literature. Kawabata was awarded the Nobel Prize in 1968. From 1968 to 1992, Kawabata remained the most frequently mentioned author. His overall peak occurred in 1974, six years after his award. Likely the reason for this peak is the amount of time it takes to create new English translations of his works as well as the time it takes to disseminate critical responses to these works. One could surmise that the rising frequency of mentioning Kawabata is partly in response to people writing about him, mulling over his new status

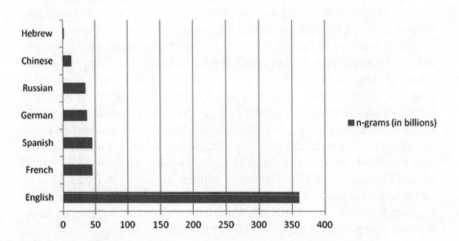

Figure 20.1. Google Ngram Viewer demonstrating frequency of *n*-grams in a variety of languages.

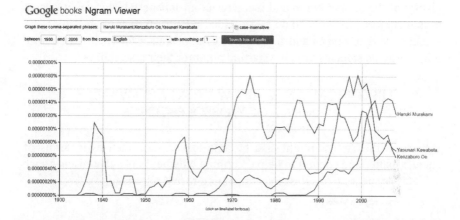

Figure 20.2. Google Ngram Viewer demonstrating frequency of *n*-grams of three well-known Japanese authors.

as Nobel laureate, the new resulting translations, and finally the scholarly and popular discussions that appear. Fame begets more fame.

A similar spike in frequency occurs for Oe as well. He was awarded the prize in 1994. At this point in time he surpasses Kawabata in the frequency of mentions until the early 2000s. His peak popularity occurred in 1998, four years after the award, and again is likely a result of the increased number of translations of his works being published. The international prize, the attendant reviews, discussions, and scholarship that arise from it appear to provide the "bump" in popularity.

Around 2003, Murakami begins to appear in the discussion as a finalist for the Nobel Prize. His popularity has increased as a result of the "Nobel bump," and the longer his name is bandied about and associated with Nobel—even if he has yet to win it—the more it propels his notoriety. The mere possibility of his winning the prize drives up the frequency of his name's appearance in print books. It should be noted, though, that Murakami's overall frequency of *n*-grams does not yet reach the historical peaks of either Kawabata or Oe.

The Ngram Viewer is truly a wonderful tool for researchers interested in the history of ideas and how events shape them. It is also a useful tool for students to get a better sense of the context of historical events and an understanding of how notoriety waxes and wanes over time. One can also track historical events in the news or the ideas that accompany them.

Finally, along with providing links to Google Books in the online public access catalog, libraries might also consider providing data visualizations of the search terms used by the library patrons to help with the frequency of terms being used. For example, say a student is interested in studying the history of the atomic bomb and is interested in Japanese history as it relates to that subject. Students could be provided a list of the terms that might be useful. The date ranges and frequencies of appearance for each search term might provide students with the starting point for their research.

HISTORICAL ANALYSIS, WAXING AND WANING TERMINOLOGY, OBSOLESCENCE

Librarians themselves might also be able to use such metrics in terms of subject matter and content to help determine whether books that discuss such subjects are nearing obsolescence or gaining recent relevance.

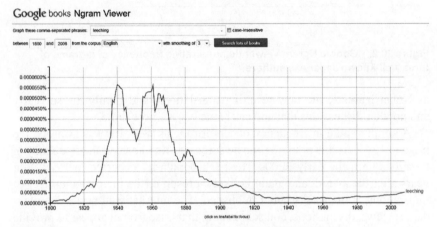

Figure 20.3.　Frequency of leeching *n*-gram from 1800 to 2008.

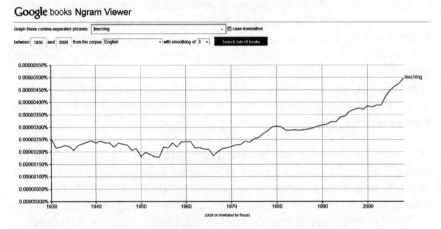

Figure 20.4.　Frequency of leeching *n*-gram from 1930 to 2008.

For example, if one takes a medical practice such as leeching (figure 20.3), a procedure that has often been portrayed by modern science as a naïve and barbaric practice in the twentieth century, one might be tempted to discard books referring to this concept. However, if one looks closely at figure 20.4, one can see an uptick in references to the term and could conclude that the library might not want to discard or weed books related to leeching. Although nowhere near the historical heights from the 1850s, one can see in the figure that mention of the term *leeching* has increased by nearly three times from its historical low in 1954 (.0000017 percent) by the year 2008 (.0000049 percent). In some circles, leeching has actually been making a comeback as a viable medical procedure for some healing purposes. The point is that libraries can make use of the terminology being used in specific disciplines to help in the collection development and weeding process.

CONCLUSION: UNLOCKING THE SECRETS OF THE DIGITAL CORPUS

The Google Ngram Viewer has great potential to improve librarianship in a number of key areas. For researchers and students, the ability to mine terms for their frequency and longevity would greatly improve their ability to more accurately find the peak and valley years associated with their topics. This would include historical figures, authors, concepts, terminology, slang usage, and so on. Pairing these *n*-gram data visualizations with search terms used in libraries (in the form of either keywords or controlled vocabularies) might help give patrons a broader perspective in their searches as well. Seeing the frequency of terms over time might improve their ability to hone their searches. Librarians might also employ such tools to help with collection development, in the broader sense, by being able to track not only the terms that are becoming obsolete but also those terms that are trending upward.

REFERENCES

Arrington, Michael. 2008. "Microsoft to Shut Live Search Books." *TechCrunch*, May 23. http://techcrunch.com/2008/05/23/microsoft-to-shut-live-search-books/ .
Authors Guild. 2013. "Remember the Orphans? Battle Lines Being Drawn in HathiTrust Appeal." June 7. http://www.authorsguild.org/advocacy/remember-the-orphans-battle-lines-being-drawn-in-hathitrust-appeal/ .
Google Books. 2014. "Google Books Library Project." http://books.google.com/googlebooks/library/ .
Jackson, Joab. 2010. "Google: 129 Million Different Books Have Been Published." *PC World*, August 6. http://www.pcworld.com/article/202803/google_129_million_different_books_have_been_published.html .
James, Ryan. 2010. "An Assessment of the Legibility of Google Books." *Journal of Access Services* 7 (4): 223–28.

James, Ryan, and Andrew Weiss. 2012. "An Assessment of Google Books' Metadata." *Journal of Library Metadata* 12 (1): 15–22.

Jeanneney, Jean-Noel. 2007. *Google and the Myth of Universal Knowledge: A View from Europe*. Chicago: University of Chicago Press. First published in French in 2005.

Michel, Jean-Baptiste, Yuan Kui Shen, Aviva Presser Aiden, Adrian Veres, Matthew K. Gray, Joseph P. Pickett, Dale Hoiberg, et al. 2011. "Quantitative Analysis of Culture Using Millions of Digitized Books." *Science* 331 (6014): 176–82.

Nunberg, Geoffrey. 2009a. "Google Books: A Metadata Train Wreck." *Language Log* (blog), August 29. http://languagelog.ldc.upenn.edu/nll/?p=1701 .

———. 2009b. "Google Books: The Metadata Mess." Presentation at the Google Book Settlement Conference, University of California, Berkeley, August 28. http://people.ischool.berkeley.edu/~nunberg/GBook/GoogBookMetadataSh.pdf .

Timberg, Scott. 2013. "Jaron Lanier: The Internet Destroyed the Middle Class." *Salon*, May 12. http://www.salon.com/2013/05/12/jaron_lanier_the_internet_destroyed_the_middle_class/ .

Weiss, Andrew, and Ryan James. 2013. "Assessing the Coverage of Hawaiian and Pacific Books in the Google Books Digitization Project." *OCLC Systems and Services* 29 (1): 13–21.

Chapter Twenty-One

Managing Electronic Resources License Agreements and Other Documentation with Google Drive

Apryl C. Price

INTRODUCTION

The growth of electronic resources in libraries is well documented. The Association of Research Libraries (ARL) statistics show that the mean total library materials expenditures for these libraries is $12,028,783, of which a total of $7,576,559 is specifically spent on electronic resource purchases. This means that 65 percent of the library materials budget is spent to gain access to electronic resources (Kyrillidou, Morris, and Roebuck 2012). The statistics also show that ongoing resource expenditures at research libraries have increased 456 percent from 1986 to 2012, and one-time resource expenditures have climbed 100 percent during the same time period (Association of Research Libraries [2012?]). The amount of content that librarians have to manage has significantly grown over the years. Whether it is an academic, public, school, or special library, the proliferation of electronic resources has made the lives of librarians very complicated. Previously, there were very few records to maintain in the print world, but that has changed greatly. Librarians now have to manage thousands of electronic journals, databases, and books. Electronic resources are not tangible like their print counterparts, so the documentation determines what is owned or leased and how it can be used.

Librarians that are responsible for managing these resources rely heavily on this documentation. They use it to decide how to provide access to the resource, to answer user questions, and to answer their own questions. Daily there are questions about the unavailability of access to electronic resources.

Therefore, librarians have to keep track of various bits of information regarding even the most insignificant journal to ensure that the library has proof that it should have access to specific volumes, issues, and years. To record this, managers of electronic resources maintain the following forms of information:

- Personal network drive space for scanned fully executed license agreements
- Original print fully executed license agreements and other order correspondence
- Publisher title lists spreadsheets
- Spreadsheet containing administrative logins for all journal, database, and e-book platforms
- Usage statistics for licensed resources
- Microsoft Outlook contacts for licensing, invoicing, and troubleshooting
- E-mails about perpetual access information, proxy configuration, account numbers, MARC records, and customer numbers
- Binder of print invoices
- ProQuest (formerly Serial Solutions) 360 products for linking URLs, holdings, and usage statistics
- Aleph Integrated Library System (ILS) for holdings and payment history
- Harrassowitz OttoEditions for payment history, access information
- EBSCOnet for payment history, access information, and customer numbers
- YBP GOBI for payment history, access information
- Coutts OASIS for payment history, access information

Most of the current commercial ILS and electronic resource management (ERM) systems are not capable of physically storing and maintaining all of this information and documentation. Libraries and commercial vendors have been building ERM systems for years, but most have not accomplished the efficiency that many librarians need. Librarians still have separate files in various locations. Deberah England (2013) discussed this issue in detail in her presentation titled "We Have Our ERM System, It's Implemented: Why Am I Still Going Here and There to Get the Information I Need?" While commercial ERM systems have continued to improve, they still lack features that would make them more attractive and effective for someone managing this deluge of information. Some libraries have built open source ERM systems with more flexibility to do what librarians need, but not all librarians and their institutions have the technology or system support to build or modify an open source ERM system, even though the system would be more efficient. Librarians are forced to find cheap, innovative, and readily available remedies to solve this problem. Cloud computing is one of the latest technological trends that seems to have significantly improved information sharing. "The cloud concept refers to programs and services that are hosted online, rather than on a local machine" (Ovadia 2013, 2). There are many popular cloud-computing services, but Google Drive is one of the best free applications.

WELCOME TO GOOGLE DRIVE

On April 24, 2012, Sundar Pichai (2012) introduced Google Drive as "a place where you can create, share, collaborate, and keep all of your stuff" on Google's *Official Blog*. Google Drive incorporates the former Google Docs suite of programs with their cloud-based storage services. Google Docs now only represents the online word processor, and Sheets is the online spread-sheet program (Lamont 2013). Cloud storage is one of the most useful fea-tures for ERM. Google Drive includes 15 GB of storage free of charge for individual users. If an institution or organization participates in Google Apps for Education or Business, then their affiliates will be offered 30 GB of free storage. Google Drive storage is shared with Gmail, so if e-mail is used on this account be aware of the amount of use. The amount of storage can be upgraded for a reasonable fee (Google 2014). Any documents, spreadsheets, or presentations created with Google's suite of programs will not count to-ward the storage limit. Depending on how Drive is used, the free storage space may be sufficient. However, prior to making the decision to use Goo-gle Drive, or any cloud storage, the privacy policy and security practices should be reviewed. These policies and practices are constantly changing, so they should be reviewed on a regular basis. To read the privacy policy for all Google apps being used, go to the Google Dashboard website (Davis 2012). The security practice Google uses is called *sharding*, and it is believed to keep data secure (Bennett and Chadha 2013). While Google Drive may not be the only location for file storage, it makes a great backup system. Google can act as backup storage for those pesky computer problems or local net-work outages that always prevent access when files are needed.

In order to use Google Drive, the user must have a Google account. If the user does not already have a Gmail account, getting an account is very easy and takes little time. An account can be created from the Google Drive website. Once the account is activated, Google Drive is ready for use. There is nothing to install to use Google Drive online. However, Google Drive can be installed on a PC or Mac to automatically sync files on a local computer for offline work. When the application is installed, it creates a local folder that syncs with Google Drive (Lamont 2013). The web interface is very efficient as well, so determine the best option based on how it will be used. Either way, content will be accessible online from any device with an Internet connection. Before working in Google Drive it is advisable to read *Instant Google Drive Starter* by Mike Procopio (2013) or *Google Drive & Docs in 30 Minutes* by Ian Lamont (2013) to learn more about the full capabilities of this innovative service.

GETTING STARTED IN GOOGLE DRIVE

Once an active account is established implementation can begin. The process is similar to the steps librarians would use to implement any new ERM system. Start by deciding what information should be included and the order it needs to be uploaded. Figure out the information that should be stored in the cloud and why. Before uploading content and cluttering the My Drive space, which is the area that houses the files and folders for the account, organize it by creating folders. Folders can easily be created by selecting the large "Create" button in Google Drive. It is a good idea to start with a file structure, but files can easily be moved in and out of folders. It may be best to organize around type of information and not publisher or vendor, which changes too often to keep up to date. Additionally, the Google Drive search enables searches by filename, partial filename, folder, file type, and keywords embedded in the document. It is easy to locate files no matter what the file structure looks like (Lamont 2013).

Documentation can be uploaded in stages to My Drive. Of course, it is best to start with the information that may be used the most or might be shared with the largest audience. License agreements tend to include all the information you need regarding an electronic resource. There generally are terms for who can access a resource and whether it will be accessed by Internet protocol (IP) authentication, proxy server, or username and password. Contracts for collections also usually include lists of the titles included in the subscription or purchase. Public services librarians and interlibrary loan staff tend to be very interested in having access to this information. Therefore, current license agreements may be the most useful to have readily available, so these can be uploaded first.

Most likely, digital files for all the current license agreements live somewhere on local personal or network drives. Publishers tend to exchange digital files with librarians in the negotiation and signing process, which is much quicker than waiting for paper copies in the mail. One issue with this constant exchange of digital agreements is that when signatures are added and the document is rescanned, the document quality can deteriorate quite a bit if scanning is not done well. It is best to check the digital files to be sure that they are not skewed and are legible prior to uploading. As Davis (2012) notes, when images with text are converted to Google Docs format, optical character recognition (OCR) is run to create text documents. However, if

Google Drive is simply being used as cloud-based storage, then the document may never be converted. Since these files are generally in PDF format, it is advisable to run OCR on the document. "A PDF created from a scanned piece of paper is inherently inaccessible because the content of the document is an image, not searchable text" (Zhou 2010, 155). Using software such as Adobe Acrobat Pro run OCR to make documents searchable prior to upload. There are also free OCR tools available that can be employed. Some scanners provide the option to make the document searchable during the scanning process as well. Documents may need to be scanned using a higher resolution than normal to get a decent searchable text document. Read Yongli Zhou's (2010) article "Are Your Digital Documents Web Friendly? Making Scanned Documents Web Accessible" in *Information Technology and Libraries* to learn more about digitizing documents. Taking the time to make sure the documents are searchable will greatly benefit use. When all the current digital license agreement files are ready, then they can be uploaded using the upload icon on the Drive website or by selecting and moving all files to the appropriate Google Drive folder on the computer hard drive.

After the files have been added to Drive, access can be granted to others who may need this information. This is another significant feature of Google Drive. Instead of being the gatekeeper for all this information and possibly being the only person to know of its existence, the ability to share pulls the documentation out of its hidden silos. Librarians and staff that engage in virtual and in-person reference, instruction, outreach, acquisitions, cataloging, and collection development can easily access the information they need to resolve patron access problems, review usage statistics, or find titles that need to be cataloged. Sending invitations to enable others to view the documentation is an important step that will help reduce the amount of time patrons and staff may normally have to wait to get the answers they seek. Instead of waiting for an e-mail or ticket to be answered, staff will be empowered to look for answers themselves. This will decrease some of the burden that managers of electronic resources normally feel to always be checking e-mail for problems and questions. Therefore, it is essential to share the documents with others in the library. Upon upload all files are set to private by default (Miller 2012). Luckily, whether the content was created in Google or not, each file has a "Share" button. Simply enter the names or e-mail addresses of those people who should be invited for viewing and the content will be shared. However, to share numerous files it is best to share the folder by making the selection from the My Drive left-side menu. Depending on the sharing setting, users may be required to have Google accounts to be able to access the documents, and those who do not will be taken to registration. Sharing visibility can be set for public viewing or limited to those with a direct URL, so that a Google account is not required for access (Miller 2012). Unless the license agreements include confidentiality clauses, this should not be a problem, but this decision will need to be made based on local practice. Additionally the owner of the content can restrict invitees from sharing the document with others within the sharing settings (Lamont 2013). This is useful for licenses with confidentiality clauses and other documentation that should not be shared widely.

Now that the active fully executed license agreements have been added to the cloud and the appropriate people have been invited to view the files, the same or similar process can be used to set up the other content. Not all resources have license agreements, so there may be copious amounts of title lists from various years to document the library's ownership or access rights. Title lists in Microsoft Excel format can be uploaded, but it may be wise to convert to Google Sheets format for easier manipulation and searching. For example, if more than one person checks and activates title access in the ERM system, then these spreadsheets should be converted to Google format. Only Google content can be edited in Drive, so conversion will allow one person to note missing title access on the spreadsheet while the other reviews the notes and activates this missing access in the ERM. Since Google Drive shares one file, the editors can see real-time changes as long as they have a working Internet connection (Lamont 2013). This is a benefit to workflow efficiency and the user experience. While the ability to edit content may be needed to manipulate title lists, not all content will need to be converted to Google format. E-mails can be saved as .txt documents and simply uploaded for sharing or as PDFs after running OCR as mentioned earlier. Instead of hunting for print invoices that may eventually be destroyed due to record laws, these invoices should also be scanned to create digital files. Many invoices include terms and conditions that the licensee will need to adhere to during the subscription or access period, so archiving this information will be just as important as license agreements. These invoices can also be used as evidence of access rights in case of publisher changes. The need to harvest usage statistics will depend on the other options for storage that are available. Most commercial ERM systems only collect COUNTER (Counting Online Usage of Networked Electronic Resources) compliant usage, and many smaller publishers and niche databases do not provide COUNTER-compliant statistics. Some usage statistics arrive in a non-COUNTER format in spreadsheets and e-mails, so these reports are a great candidate for addition to Google Drive. Faced with flat budgets and declining buying power, librarians and staff are paying more attention to usage statistics. With shared access to this content, the manager of electronic resources can spend more time on the important tasks and worry less about responding to the endless requests for terms of use, title lists, and usage statistics.

MANAGING MY DRIVE

Now that Google Drive has been supplied with all the intended content and shared with the appropriate people, the system will need regular curation. When licenses are retired, it may be best to move them from the cloud to a local network drive. New licenses can be uploaded and converted to Google Docs format for external review by general counsel, purchasing, or accounting and then collaborative edits can easily be made prior to signing the agreement. Title lists will need to be added every year to stay up to date, so remembering to gather them will be important. Perpetual access rights depend hugely on knowing the years paid, so maintaining these lists for all years is a necessity unless there are no rights to ensure. Of course, invoices and e-mail history can document this as well. It is a good idea to upload the e-mail messages as soon as possible, because e-mail tends to get lost or forgotten among the many other messages. The usage statistics may require more maintenance, as it may not be necessary to maintain this data forever. Storing three years of usage may be sufficient, but that decision will have to be made to align with institutional needs.

Everything, or almost everything, is in the cloud now, so this is where the fun begins. In meetings, when asked about the usage statistics or license terms for a database, the answer can be researched immediately, either by accessing a computer, tablet, or smartphone. The web-based version of Google Drive has been discussed, but there are also apps for mobile use. It may be more difficult to view large documents, but it is possible. Android and Apple iOS have wonderful apps that allow all the content to be easily accessed. Even if there is no Internet connection, files can be kept on the device for offline viewing. This option can be toggled on or off through the app. This is useful for responding to staff and patron questions while traveling. The Google Drive app in conjunction with the Adobe Reader app allows fully executed documents to be accessed, read, and searched. This is why the OCR process is essential for scanned PDFs; these documents cannot be keyword searched without it. When the file is searchable, "interlibrary loan" can be searched to locate that section of terms within the license agreement. This is an especially helpful feature when trying to view documents on the small screen of a tablet or phone. Google is always releasing upgrades to the web and mobile interface of Drive, so the impressive capabilities and usability will only improve.

CONCLUSION

Electronic resources have been on the rise in all libraries for a while, but libraries still struggle to manage them effectively. Commercial companies have been building ERM systems for years, but they still lack some of the features that librarians need. Open source systems have greatly complemented the ERM system offerings, but if there is little or no technology support available, they are impossible to manage. Faced with these issues, free cloud-based services have become very attractive. Although privacy and security concerns are a definite challenge of cloud computing, access to content and services across devices makes it extremely appealing. While there are other free cloud-based services that could be used in a similar manner to manage ERM, Google Drive provides ease of use that is not quite present in other services. The ability to store license agreements and other documentation in the cloud so that other staff can easily view them reduces some of the burden the manager of electronic resources may feel as the gatekeeper of this information. Although giving the staff access to this information will relieve some of the burden, the manager for electronic resources will still be tasked with answering questions. Sharing the license terms and title lists with library staff gives them the opportunity to attempt to research questions and resolve access issues on their own, instead of waiting for the gatekeeper. Resolving issues quicker will improve the user experience, which is why proper management of ERM is so important. The future of cloud-based computing is unknown, but why not experiment and see which tools provide the desired efficiency. The technology is always evolving, so take a risk and do things differently.

REFERENCES

Association of Research Libraries. [2012?]. "Expenditure Trends in ARL Libraries, 1986–2012." Washington, DC: Association of Research Libraries. http://www.arl.org/storage/documents/expenditure-trends.pdf.

Bennett, Frank, and Peter Chadha. 2013. *Thinking of . . . Going Google Apps? Ask the Smart Questions*. Hampshire, UK: Smart Questions Limited.

Davis, Yvette. 2012. *Google Secrets: Do What You Never Thought Possible with Google*. Indianapolis, IN: John Wiley.

England, Deberah. 2013. "We Have Our ERM System, It's Implemented: Why Am I Still Going Here and There to Get the Information I Need?" *The Serials Librarian* 64 (1–4): 111–17. http://dx.doi.org/10.1080/0361526X.2013.760148.

Google. 2014. "Storage Limits." *Google Apps Help Center*. https://support.google.com/a/answer/1186436?hl=en.

Kyrillidou, Martha, Shaneka Morris, and Gary Roebuck, eds. 2012. *ARL Statistics 2010–2011*. Washington, DC: Association of Research Libraries. http://publications.arl.org/ARL-Statistics-2010-2011/.

Lamont, Ian. 2013. *Google Drive & Docs in 30 Minutes: The Unofficial Guide to Google's Free Online Office and Storage Suite*. [Massachusetts]: i30 Media Corporation.

Miller, Michael. 2012. *Introduction to Google Apps: Productivity Apps*. Boston: Pearson Education.

Ovadia, Steven. 2013. *The Librarian's Guide to Academic Research in the Cloud*. Oxford: Chandos Publishing.

Pichai, Sundar. 2012. "Introducing Google Drive . . . Yes, Really." *Google Official Blog*. http://googleblog.blogspot.com/2012/04/introducing-google-drive-yes-really.html.

Procopio, Mike. 2013. *Instant Google Drive Starter*. Birmingham, UK: Packt Publishing.

Zhou, Yongli. 2010. "Are Your Digital Documents Web Friendly? Making Scanned Documents Web Accessible." *Information Technology and Libraries* 29 (3): 151–60. doi:10.6017/ital.v29i3.3140.

Chapter Twenty-Two

Simplifying "Contact Us"

How to Offer Free Multichannel Virtual Reference and Monitor It from a Single Unified Dashboard

Laura Baker

With the prevalence of mobile devices, many libraries recognize the value of providing multiple ways for patrons to contact the library. Chat, e-mail, and text messaging as well as traditional telephone are now standard and expected communication channels (Chow and Croxton 2012; Zickuhr, Rainie, and Purcell 2013). Some libraries have tried do-it-yourself methods by signing up for a variety of separate e-mail and texting services only to find they have to log in to and monitor a dizzying array of unconnected accounts. This quickly becomes too much to manage. Other services offer a more integrated approach, but they often come with a fee (Thomsett-Scott 2013; Tisi et al. 2014). By using services from Google and Google Labs, a library can offer free texting, e-mail, phone, and voicemail and then tie all these communication channels into one account for easy and efficient monitoring. The result is less frustrated staff, better service to patrons, and a leaner bottom line for the library's budget.

See figure 22.1 for a sneak preview of what this chapter will help you create for your library.

Figure 22.1 shows a full-screen view of the inbox from our library's Google account that we use for virtual reference. You can see an enlarged version of the complete screen online (Baker 2014) as well as in the accompanying screenshots in this chapter. Figure 22.1 shows the layout of the entire inbox. This is what we see when we log in. Figure 22.2 and figure 22.3 show enlarged and more legible views of the left and right halves of the inbox.

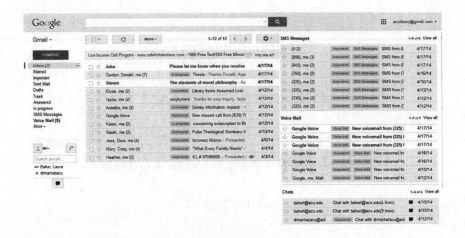

Figure 22.1.   Full-screen view of the inbox.

The inbox works the same way as any other inbox, but it has some special features added. Notice that the inbox is divided into separate panels for the types of messages coming into it. E-mail messages are on the left panel, which has been zoomed in for figure 22.2. The right side of the inbox, shown in figure 22.3, has places for SMS messages (text messages), for voicemail left through phone calls, and for transcripts of live chats. In all cases, new messages are in bold to alert us to their arrival. Other messages are marked as to whether they have been answered or are in progress. Everything has a folder in the far left column under "Inbox," where we can file messages for our records later and keep the inbox neat.

Each person responsible for providing reference service logs in to this one unified account. He or she optionally starts a separate chat account, which I will explain later. From this one interface, we can see at a glance all the various ways patrons might use to contact us. The messages come to this central account instead of requiring us to check multiple locations. With this method, we have a simpler way of managing questions at a more budget-friendly price.

Here is how to do it for your library.

STEP ONE: SIGN UP FOR GOOGLE MAIL AND GOOGLE VOICE

The library should sign up for a Google account if it does not already have one. A single Google account gives access to a host of Google services, so if the library already uses Gmail, YouTube, Picasa, and so on, it already has an account in place.

				1–12 of 12	

Low Income Cell Program - www.safelinkwireless.com - 1000 Free Text/250 Free Minute ⋯ Why this ad?

		John	Please let me know when you receive	4/17/14
	☆	Gordon, Donald, me (7)	In progress Thesis - Thanks Donald. Appi	4/17/14
	☆	Steven	The elements of moral philosophy - An	4/17/14
	☆	Elyse, me (2)	Answered Library Items Assumed Lost -	4/12/14
	☆	Taylor, me (2)	emplyment - Thanks for your inquiry, Taylo	4/12/14
	☆	Annetta, me (2)	Answered Dewey information request - I	4/12/14
	☆	Google Voice	Answered New missed call from (830) 7(	4/12/14
	☆	Karen, me (2)	In progress concerning subscription to Br	4/12/14
	☆	Sarah, me (2)	Answered Fuller Theological Seminary Ir	4/10/14
	☆	Jess, Dora, me (4)	Answered Incorrect Notice - Forwarded	4/6/14
	☆	Mary, Craig, me (4)	Answered "What Every Family Needs" -	4/4/14
	☆	Heather, me (2)	Answered ILL # 97086908 - Forwarded r	4/3/14

Figure 22.2. Enlarged view of the left side of the inbox.

Go to Google and click the blue Sign In button in the top-right corner. Sign in with your library's Google account. If the library does not already have a Google account, you can create one here. It is helpful to have an account name that makes sense to users, such as aculibrary@gmail.com or askyourlibrary@gmail.com. If multiple people will be responsible for monitoring virtual reference, set a password that can be securely shared with everyone in that group.

This suite of library services will use Google Voice as its foundation. Registering for Google Voice lets the library have a free number that can be used for telephoning and texting. Use the Google Apps button in the top right to go to Google Voice. You may need to click the More button several times to display Google Voice in the list of apps.

Once in Google Voice, select Get a Voice Number and then click I Want a New Number. Follow the prompts. The library can specify which phone it wants the Google number to ring, such as the research desk or the circulation desk. Google automatically sets up texting with your Google Voice number, too. When finished with the Google Voice registration, the library will have one Google number that patrons can use either to call the library at its regular telephone number or to text the library. In turn, the library can send and receive text messages for free. It will also have a free Google e-mail, or Gmail, account through the same service.

2	<	>	⚙ ▾	SMS Messages				1–8 of 8 View all
				☆ (512)	Answered	SMS Messages	SMS from (5	4/17/14
Free Minute ⋯ Why this ad?				☆ (806), me (3)	Answered	SMS Messages	SMS from (8	4/17/14
u receive		4/17/14		☆ (940), me (2)	Answered	SMS Messages	SMS from (9	4/17/14
onald. App		4/17/14		☆ (760), me (2)	Answered	SMS Messages	SMS from (7	4/16/14
sophy - An		4/17/14		☆ (325), me (2)	Answered	SMS Messages	SMS from (3	4/16/14
med Lost -		4/12/14		☆ (325), me (2)	Answered	SMS Messages	SMS from (3	4/12/14
quiry, Taylo		4/12/14		☆ (330), me (2)	Answered	SMS Messages	SMS from (3	4/12/14
request - f		4/12/14		☆ (325), me (2)	Answered	SMS Messages	SMS from (3	4/12/14
om (830) 7(		4/12/14						
iption to Br		4/12/14		**Voice Mail**				1–7 of 7 View all
Seminary Ir		4/10/14		☆ **Google Voice**	Voice Mail	**New voicemail from (325) :**		4/17/14
Forwarded		4/6/14		☆ **Google Voice**	Voice Mail	**New voicemail from (325) (**		4/17/14
y Needs" -		4/4/14		☆ **Google Voice**	Voice Mail	**New voicemail from (325) (**		4/17/14
Forwarded r ∞		4/3/14		☆ Google Voice	Answered	Voice Mail	New voicemail frc	4/16/14
				☆ Google Voice	Answered	Voice Mail	New voicemail frc	4/16/14
				☆ Google Voice	Answered	Voice Mail	New voicemail frc	4/15/14
				☆ Google, me, Mail	Answered	Voice Mail	New voicemail frc	4/12/14

				Chats				1–3 of 3 View all
				☆ bakerl@acu.edu		Chat with bakerl@acu.edu(4 lines)	💬	4/15/14
				☆ bakerl@acu.edu		Chat with bakerl@acu.edu(9 lines)	💬	4/15/14
				☆ drmarkatacu@aol.	Answered	Chat with drmarkatacu@aol.	💬	4/12/14

Figure 22.3.   Enlarged view of the right side of the inbox.

STEP TWO: ROUTE MESSAGES

A significant power of Google Voice is the ability to tie all incoming messages together in one place for easy viewing, regardless of whether they are SMS text messages, voicemail, or e-mail. You do not want to have to look in one place to check e-mail, another to answer a text message, pick up the phone for voicemail, and so forth. With a couple of settings, all of the library's messages can be funneled into one account, so you can see everything at glance and respond accordingly.

To route text messages to the library's Gmail:

1. Enter the library's Google Voice account.
2. Click Settings (gear icon) near the top right.
3. Select Settings, then the Voice Mail & Text tab.
4. Under Text Forwarding, check the box to Forward Text Messages to My Email.

Now when someone sends the library a text message, it will show up as an e-mail message in the library's Gmail inbox. This is a wonderful feature. You do not have to have a "common mobile phone" that everyone tries to share for reference. You do not have keep an extra device like a smartphone nearby. When text messages come in, you can type on a regular-size keyboard and respond using e-mail that is already familiar. The response will be sent to the patron via SMS text message, the same way it came in.

Routing Voicemail

If you route voicemail, any telephone messages someone leaves when they call the library using the library's Google number will be forwarded to the Gmail account. This is handy for phone calls when the librarian is on another line or when the library is closed. It applies only to phone calls that were dialed using the Google number, not the library's regular telephone number (a good case for using the Google number as your primary contact).

1. Google Voice—Settings Gear icon—Settings—Voicemail & Text tab
2. Under Voicemail Notifications, check Email the Message To and fill in the library's Gmail address.
3. Under Voicemail Transcripts, check Transcribe Voicemails to get a rough, written translation of the spoken voicemail. See the section below on creating the dashboard for a way to listen to the recording as well.

Forwarding other E-mail to Gmail

Your library may have an official group e-mail account through your city or institution, something that does not end in gmail.com. It is extremely helpful to set that account to forward messages to the library's Gmail account for virtual reference, otherwise you will have to check both accounts for possible e-mail messages. The easiest way to forward e-mail is to check your account settings on your institutional e-mail account, asking your e-mail provider if necessary. Many accounts have a way to enable forwarding very easily. Google also has a helpful procedure for more information (Google 2014).

STEP THREE: CREATING THE DASHBOARD

All of the previous steps route messages to the library's Gmail inbox. If we stopped here, e-mail, voicemail, and text messages would be intermingled in one long list. Creating a dashboard lets us divide the inbox into sections that can organize messages by type. This makes for a display that is cleaner and easier to use.

To create the dashboard, we will use Google Labs. Google Labs are extra features you can enable to extend the functionality of Gmail. We'll turn on a couple of features and then set the messages to filter automatically into the appropriate category.

1. Library Gmail—Settings Gear icon—Settings—Labs tab
2. Locate the feature called "Multiple Inboxes." You can scroll down or use the search box to find it. Click the radio button to enable. This extension divides the inbox into sections that we will label shortly.
3. You may also want to enable another Lab called "Google Voice Player in Mail." This will let you play a recording of any voicemail messages left in your inbox. The inbox will already show a written transcript of the message, but it may be unreadable, especially if the person was not speaking clearly. For that reason, I highly recommend enabling Google Voice Player in Mail.
4. Click Save Changes near the bottom when you are done.

Table 22.1.   Table Settings for the Screen Displayed in Figure 22.1

	Search query	Panel title (optional)
Pane 0:	label:sms-messages	SMS Messages
Pane 1:	label:voicemail	Voicemail
Pane 2:	is:chat	Chats

Go back to Gmail Settings and click it again to refresh the account. You should now see a new tab called "Multiple Inboxes." Click this tab. Here is where you specify the behavior and look of the partitioned inbox.

The main inbox already gathers e-mail and does not have to be defined. We need to create separate panels for SMS messages, voicemail, and anything else you wish. The column for "Search Query" lists a label that we will apply to messages so Google can determine what kind of message it is and sort it appropriately. The column for "Panel Title" is where you type the header you want to appear over that section. The settings we used for the inbox pictured in figure 22.1 earlier in this chapter are provided in table 22.1. Other settings let you specify how many messages to show in each of the panes and where you want them to appear on the screen. Ours are positioned to the right of the inbox.

The final step is to establish filters to match the labels we specified under "Search query."

1. Gmail—Settings Gear icon—Settings—Filter tab—Create a New Filter
2. Fill in the following:

   a. From: text.voice@Google.com.

  b. Click Create Filter with This Search for the next screen.
  c. Check Skip the Inbox (Archive it).
  d. Check Apply the Label: SMS Messages (you may have to click New Label to type this).
  e. Click the Create Filter button to save.

And for voicemail:

  1. Create a new filter.

      a. From: voice-noreply@Google.com.
      b. Subject: voicemail.
      c. Click Create Filter with the Search for the next screen.
      d. Check Skip the Inbox (Archive it).
      e. Check Apply the Label: Voicemail (you may have to click New Label to type this).
      f. Click the Create Filter button to save.

You are free to use whatever you want for the labels. The only requirement is that the label you create in the filters must match a hyphenated version of the label you specify in the "Multiple Inboxes Search Query" column.

Chat Options

In my example, we have an extra panel set up for chat. This automatically uses Google Chat, which many people have and use. Unfortunately, a patron must already have a Google account to use Google Chat. Since our campus issues Gmail accounts to all students and campus employees, many of our patrons are naturally plugged into Google Chat, and it made sense for us to incorporate that channel into our dashboard. However, we also wanted a way for anyone to be able to chat with us through a widget we could embed on our library website, something that did not require them to have an account with a certain company. We therefore found it necessary to go with an additional separate service that offered such an open chat platform. We reviewed chat tools that fit our criteria for embeddable widgets that would work with any user ("Meebo Messenger" 2012). Since the demise of the popular Meebo, we have since started using Zoho as our supplementary chat. Zoho does not yet integrate with Google, so we sign on to our Zoho account in addition to our Google dashboard. Zoho runs in the background, alerting us with an audible signal and a pop-up box if a live chat is coming in. It is the one exception to our otherwise totally unified dashboard, but the nature of chat and the benefits of the other Google services do not make the necessity of a separate chat login seem onerous. Your library may go with Google

Chat, supplement it with another service as we have with Zoho, or omit it entirely as your community needs suggest.

Now you have an inbox that collects not only your e-mail messages but a lot more. It acts as a central communication center, providing services as well as a one-stop place to receive and respond to messages of all types. You can see questions by e-mail, send and receive text messages, read or listen to voicemail from the Google phone after hours, and all within the same Google interface. This one account replaces a host of devices and separate, fee-based services and does it for free. The virtual reference desk just got simpler.

STAFF TRAINING AND TIPS

Using the unified dashboard is very simple, especially if staff is already familiar with e-mail in general. Sometimes, though, training is less about teaching new skills than managing comfort levels. No technology project is complete without introducing staff to change that affects them and making sure everyone is on board. Training is also a good time to establish some policies for handling the messages so that everyone who logs in follows the same procedures.

Beta test before roll-out. Ask a couple of librarians or staff if they will try out the service with you by exchanging messages on the central account. If you have key opinion setters in your library or staff who might be unsure of change, then involve them in the test group. Most people are flattered to help and welcome the opportunity to try something out beforehand. Be sure to take their suggestions on how the inbox panels are labeled, the color of the interface, the position of the panels, and any other aesthetic or operational ideas they have.

Once you are satisfied with the beta version, demonstrate the account to the whole staff. Log on using a computer connected to a projection system so everyone in the room can see the screen. Tell everyone the group login and password and that this information should be kept secure, just like that for any other library database.

Role-play by asking staff to send a text message or a phone message as if they were a patron. Let them see how those messages appear and what the librarian on duty does to respond. Seeing the communication from both sides of the exchange helps everyone feel comfortable with the procedure and know what the patron will see as a result.

Use tags to mark the status of the messages. Since everyone will be looking at the same account, there needs to be a way of knowing if a patron's question has been answered, if someone else is working on it, or if it has been received but waiting for attention. We use the Gmail Labels feature to tag messages as "in progress" to show that the question has been assigned to someone in the library who is working on an answer. We use the label "answered" when a reply has been sent and to mark questions as finished. We then move the answered questions to a Gmail folder to keep the inbox clean.

MAXIMIZING THE BENEFITS

Using Google for virtual reference service offers a number of other features that a library can use to its advantage. These were things that our library did not expect but that we discovered, to our delight, after using it for a while.

- *Reference backup.* When we set up Google Voice, we specified that the Google Voice number should ring the telephone at the library's reference desk. In fact, you can specify up to six different locations to which Google Voice will forward calls. We use this as a telephone tree so that the phone will ring alternative places if the reference desk phone is busy or if the person covering the desk needs to be in a different location. It is also a handy way for our front desk to reach a librarian during off-peak hours when we are not staffing the reference desk. Instead of trying various office numbers to find someone who is in, the front desk calls the Google number and lets the call forwarding ring up to six people who are on backup that day.
- *Untethered reference desk staffing.* Remember that the Google dashboard can be accessed anywhere from any device connected to the Internet, including a computer tablet. With tablet in hand and a cell phone that the Google number is set to ring, the librarian is free to get out from behind the desk, walk among the patrons, or even work from another building or from home and still be able to respond to virtual reference questions. Questions go to a person, not to a physical location. This is the core of virtual and roaming reference.
- *" One-Number Access."* With the method described here, the same phone number patrons use to telephone the library is the number they use to text the library. Through marketing, the library can capitalize on this convenience to patrons, making the library contact information easy to remember and easy to use.
- *Built-in knowledge bank.* Google's inbox has a convenient search function that lets one search messages by date, keyword, label tags, and a number of other characteristics. Since Google automatically archives messages unless you delete them, you can search for answers to previously asked questions and save effort instead of starting from scratch. This makes the dashboard act as a knowledge bank that can document hard-to-find an-

swers. A knowledge bank is a premium feature of many fee-based services, but it a free by-product here.

- *Easy statistics via labels*. Proper use of labels lets you gather reference statistics more easily later on. Google automatically tags some messages by type, such as "SMS message" or "voicemail." As previously mentioned, we also apply labels ourselves, such as "answered." Messages may have more than one label. By combining labels with the search function, we can display messages we need for our statistics. To see how many text messages we received for the month, we simply click the SMS Message folder where all these are stored. To see how many questions we answered virtually for the semester, we search by date for questions labeled "answered." Google displays the count along with the questions and our replies. The library can get analytics on questions by type, by date or time, by answered/unanswered status, or by any combination that labels and the advanced search make possible.

Every library must use its limited resources to the fullest, especially during tough economic times. It behooves each library to consider all of the choices of software services it can and not just the fee-based products from popular vendors. Virtual reference is something each library must consider, and Google services make it possible for all of us, regardless of our size. By employing the tips described in this chapter, you can build your own virtual reference suite of services and customize it for your needs, turning your reference inbox into an easy-to-use communication center and making your dollars go that much further.

REFERENCES

Baker, Laura. 2014. *Unified Inbox*. bit.ly/UnifiedInbox.
Chow, Anthony S., and Rebecca A. Croxton. 2012. "Information-Seeking Behavior and Reference Medium Preferences." *Reference & User Services Quarterly* 51 (3): 246–62.
Google. 2014. "3 Ways to Move Email from Other Accounts to Gmail." *Gmail Help Forum*. https://support.google.com/mail/answer/56283?hl=en.
"Meebo Messenger Is Shutting Down, Now What?" 2012. *Lassana Magassa*. https://web.archive.org/web/20131225130835/http://www.lassanamagassa.com/2012/06/meebo-messenger-is-shutting-down-now-what/.
Thomsett-Scott, Beth. 2013. "Virtual Reference Services: Considerations and Technologies." In *Implementing Virtual Reference Services: A LITA Guide*. Chicago: ALA TechSource.
Tisi, Madel, Anthony C. Joachim, Maria Deptula, Denise A. Brush, and Michelle Martin. 2014. "The Power Punch of Virtual Reference Tools: LibraryH3lp, LibAnswers and Mosio." Presented at the VALE New Jersey Association of College and Research Libraries 2014 Annual Users' Conference, Piscataway, NJ, January 10. http://www.valenj.org/conference/session/b15-power-punch-virtual-reference-tools-libraryh3lp-libanswers-and-mosio.
Zickuhr, Kathryn, Lee Rainie, and Kristen Purcell. 2013. *Library Services in the Digital Age*. Washington, DC: Pew Internet and American Life Project. http://libraries.pewinternet.org/files/legacy-pdf/PIP_Library%20services_Report.pdf.

Chapter Twenty-Three

Does Google Scholar Help or Hurt Institutional Repositories?

Peace Ossom Williamson and Rafia Mirza

Librarians often act as though Google Scholar is our new frenemy; however, the reality is somewhat more complicated. While it is true that Google Scholar's requirements are not always transparent to us (us being libraries, archives, institutional repositories), enough research has been conducted that we can now make an educated guess as to how to organize the content in institutional repositories (IRs) and archives in ways that Google Scholar can index more effectively. Google Scholar offers many options for libraries to make their IR collections discoverable. It is true that there has been a history of Google Scholar lacking in transparency and structure; what's more, Google Scholar citations profiles are largely accomplishing the same objectives as IRs. This chapter looks at the effect these Google programs have on IRs and how libraries can respond.

Google Scholar has gained its powerful position because libraries have yet to create an effective means of searching across IRs. The combination between this lack of otherwise-created searching capability and the large general public preference toward Google resulted in Google Scholar becoming the default search mechanism for scholarly works across a broad range of locations. Librarians' learning to navigate in Google's territory arises from the growing movement toward open access content, especially in scholarly work. This open access movement arose from the ongoing serials and access crisis, where the costs of serials are increasing rapidly. Most libraries' budgets cannot accommodate these cost increases, and they must cut access to valuable scholarly works. Now, only the more affluent—typically Western—institutions can afford the most expensive resources. As scholars continue to write for impact rather than recompense, publishers profit handsomely from scholarly works.

Open access differs from this traditional structure, as scholarly works are provided without charge to readers, lessening the purchasing pressure on libraries. Publishers are slow and resistant to provide open access to their publications, and, in response, libraries are storing the works of their affiliated scholars in repositories in order to provide ease of access and archiving of these works. These repositories are institutional, as the resources present are works of individuals from that particular institution. IRs contain the metadata and full text of scholarly work produced by people affiliated with that institution. Works that are otherwise hidden behind a paywall, like articles in medical journals, or works that are largely inaccessible to the greater public, like theses and dissertations, are made available in order to increase the impact of work stemming from a particular institution. IRs include a wider range of resources than traditional journals or publications. Student work, gray literature, and academic and professional presentations are also included.

While there are major differences in the more than three thousand repositories existing across the globe, IRs provide research to the greater community through the green open access model—that is, there is neither a cost to deposit works nor a cost to access works. Green open access puts control of publication in the hands of the research community and can improve dissemination, if done well. This increases the number of citations, as these resources are no longer limited to users affiliated with Western institutions with greater amounts of financial capital. In response, publishers are beginning to offer gold open access models. Gold open access models are models in which authors may pay some of the costs of publication, but access to full text within these publications have no cost. While somewhat beneficial, IRs have a greater benefit in that they cut publishers' ability to control pricing, return control of research to the hands of the researcher, and improve resource "sustainability and stewardship" through the management and preservation of original output (Poynder 2014).

INDEXING AND DISCOVERY

Google consistently has more than 65 percent of the share in search engine use, and this rate has also been true of Google Scholar (Arlitsch and O'Brien 2012). Because of Google Scholar's dominance, many users are able to find useful literature from IRs without any prior knowledge of the purpose or existence of these repositories. It also provides about three-fourths of traffic to IRs (Poynder 2014). Furthermore, it provides a proven location for centralized searching and a familiar interface for most users. The indexing of IRs by Google Scholar is of importance due to their relationship. Google Scholar is a search engine independent of Google Search in that its purpose is finding scholarly works regardless of where they are located on the web. In that pursuit, "Google privileges information from reputable sites, such as those with '.gov' or '.edu' or 'ac.uk' domain names" (Dawson and Hamilton 2006). Google Scholar has been heralded in its ability to crawl publisher pages, open access databases, IRs and other repositories, and author websites in addition to other locations for research literature. Although designed specifically for simple searching, it is an outstanding resource for obtaining gray literature (Giustini and Kamel Boulos 2013). In fact, since Google Scholar shows the primary version of the article (which is generally not the preprint in the IR, but the post-print at the publisher's site), it may be better for IRs to focus on gray literature (Arlitsch and O'Brien 2012).

Because Google Scholar is the most preferred method of searching for scholarly works and, therefore, the primary indexing service (even if by default), IR administrators must be aware of its indexing methods. In order to increase discoverability, IRs need to take into account what information Google Scholar is using for indexing and how it interprets our metadata. If we do what it wants and accommodate its needs, it is great for our IRs. If we do not, then it can greatly diminish our discoverability by making our IRs almost completely invisible. The major setback is that librarians have to guess how this is done. Google is secretive about its algorithm, minimizing users' ability to perform expert searches. It has advanced searching capabilities; however, it has very few limiters and no controlled vocabulary. Nevertheless, Dawson and Hamilton (2006) point out that some sites, such as Physics Finder, have optimized their pages to get higher page rankings, while keeping in accordance with their metadata schema.

When it comes to the public interface, it is also difficult for users to determine which materials fall within or outside of Google Scholar's scope. Google and Google Scholar indexes are separately created and maintained due to Google Scholar's focus on peer-reviewed articles, books, white papers, patents, and legal reports; therefore, results vary vastly when the same search is completed within Google or within Google Scholar. Results are organized by key words, the number of citations, and the number of previous clicks on the links; therefore, many users struggle to retrieve newly published materials or relevant research. Searching for titles or author names is a much more successful method of using Google Scholar, while topic and keyword searches remain largely unpredictable. There are ways to increase the rankings of articles in repositories by including information such as subject headings.

OPTIMIZING DISCOVERABILITY

The indexing ratio for an IR in Google Scholar is the sum of the IR's unique URLs in Google Scholar over the total sum of unique URLs in the repository. This ratio is typically low in the average IR due to the conflict between Google Scholar metadata requirements and the procedures of many IRs. Search engines are severely restricted to searching text, and they cannot read text present within multimedia, JavaScript, or images; moreover, IR databases and servers must allow search engine crawlers to be present in the first place. The crawlers then follow links to the metadata, which is evaluated by Google Scholar algorithms, and these algorithms determine whether that information is added to the Google Scholar index (Arlitsch and O'Brien 2012, 64). Google Scholar provides guidelines for search engine optimization (SEO) on its site through improving the success of its crawling and indexing websites and repositories. Some of these guidelines recommend using up-to-date software and providing chronological lists of works and permanent links (Google Scholar 2010). Most notably, Google Scholar (2010) discourages the use of Dublin Core as a metadata schema. Arlitsch and O'Brien found that IRs that adhered to the indexing guidelines put forth by Google Scholar had an indexing ratio of 88 percent to 98 percent, while IRs that did not had a much lower indexing ratio of 38 percent to 48 percent (2012, 70).

The question then, of course, is, should librarians adopt wholesale a metadata schema that is not created by librarians? Should SEO for Google Scholar be a primary concern of ours? The short answer to these questions is a resounding, "Yes!" To a certain extent, we need to accept *satisficing* as the primary searching method of the average user; therefore, we cannot dictate how users come to our online repositories. The goal of an IR is to make the content more accessible; the goal is not necessarily prominence or recognition for the IR itself. In addition, Dublin Core works poorly as a schema for articles because it is open to inconsistent interpretation in practice. For example, Dublin Core includes one field for multiple citation data, including journal name and volume number, and there are no fields distinguishing document type (Arlitsch and O'Brien 2012). However, as is often said, "Metadata is a love note to the future," and we want to make sure IRs include the metadata that future librarians and archivists may find useful in addition to adapting to the ways our current users find resources in our IRs. We can do both by using current standards, such as Dublin Core, and by also adding whatever code or script overlays are needed for the materials in our IRs to be indexed by Google Scholar. We want to do more than merely pay lip service to interoperability, yet we do not want to go overboard in allowing Google Scholar to completely dictate our choice of metadata schema, as there is no guarantee that Google will always be the Internet's search engine of choice.

We want to prioritize future searchers as much as we prioritize current searchers.

ADAPTING TO METADATA SCHEMAS

Libraries must adapt to the online environment by updating access and assessment of resources in the IR. For example, Google Scholar gives users a direct link to the PDF full text of resources found, thereby bypassing any context. This circumvents the libraries' mechanisms used in keeping visitation statistics for PDFs that are separate from the HTML display. In response, libraries should add a PHP script to more accurately track usage statistics (Arlitsch and O'Brien 2012). Libraries can also use schemas—for example, PRISM or Highwire Press—that are better able to accept citation information. Furthermore, it is of utmost importance that IR administrators avoid errors including dead links or slow servers because when crawlers encounter errors on a site, they are less likely to return. Google Webmaster Tools in combination with assessment tools can reduce instances of errors and improve analysis of the impact of an IR on its institution.

Dawson and Hamilton use the term *data shoogling* to "refer to the process of rejigging, or republishing, existing digital collections, and their associated metadata, for the specific purpose of making them more easily retrievable via Google" (2006, 313). The four standards they espouse are:

1. Search engine optimization
2. Metadata cleaning
3. Metadata optimization
4. Metadata exporting

Basically, they argue that we should engage in consistent metadata, keeping in mind the constraints of the web page and what we do know about what web crawlers look for. This seems a reasonable argument. Dawson and Hamilton discuss the ways in which libraries can export their collection's metadata to a series of static pages (that get updated to reflect the collection if it changes) that get indexed in Google Scholar, increasing the likelihood that people will find the collections using subject terms. They give the example of the Glasgow Digital Libraries, whose collections received high ranking in Google Scholar without any external links, seemingly simply by having "search-engine-friendly design" (Dawson and Hamilton 2006, 319). If librarians include Library of Congress subject headings (LCSH) in the pages for their collections, it aids searchers because those LCSHs are phrases: if a searcher uses even part of that LCSH phrase in their search, they are more likely to find relevant content. Indeed, in the example of the Glasgow Digital

Libraries, users were finding content by searching generally on a topic, not by specifically searching for the holdings of the Glasgow Digital Libraries. Arlitsch and O'Brien (2012) found that if you expressed IRs' Dublin Core metadata schemas in HTML meta tags, those IRs had higher indexing rates. Thinking of search engines as "users with substantial restraints" in regard to what types of content they can view (multimedia, JavaScript, etc.) may be the best way for librarians to think about these indexing services.

Google Scholar is how many people find open access academic resources. As librarians increase the promotion of open access content, we should think about ways in which we can maintain metadata standards, but also be receptive to the ways in which search engine crawlers harvest sites such as IRs. While Google Scholar does not have a metadata HTML tag exclusively for LCSHs, if librarians include LCSHs in their HTML meta tags or titles, they will improve their sites' rankings.

ATTRACTING ORIGINAL CONTENT

In addition to discoverability, librarians must also be aware of the robustness of their IR. Many authors have yet to be convinced of the benefits of depositing their works in IRs. This is a difficult issue to overcome, which is why IRs consist of few primary articles. It becomes a burden with no benefit in situations where academics are already publishing in open access journals.

Low percentages of primary articles have a great effect on IR discoverability because IRs that provide a larger number of primary articles, especially gray literature, are more likely to be included in Google Scholar results and rank higher in these lists, thereby improving their indexing ratio (Arlitsch and O'Brien 2012). As indexing ratios—a measure of IR discoverability—increases, scholars are more likely to deposit their original work or primary articles. Attempting to change one or both of these two codependent characteristics can create a difficult cycle for libraries. In order for libraries to break out of this cycle, they must first convince academics to deposit their already-published works in the IR then convince those same academics that, from that point forward, they need to negotiate with publishers after the peer-review process for rights to the final edited copy of their works. Methods of convincing students and faculty members to deposit their works in an IR include the following benefits:

- An increase in their works' citation counts and potential impact
- The retention and preservation of published works
- The author's access to his own works, even if the institution's subscription is canceled

- The retention of the author's copyrights and use of one's own work in teaching and in self-promotion
- The reduction in plagiarism, since the original source is openly available for referral

A common response from academics is that they are already using the Google Scholar citations profile or a similar profile via other websites, such as Academia.edu . Google Scholar profiles were made available in 2011, and users are able to create a profile by selecting their publications in Google Scholar. The individual's publications are subsequently listed, and that user is provided with citation metrics and graphs.

It is difficult to know what citations are included in these indexes because the list of resources used in Google Scholar is privately kept. Google Scholar citation information can, in ways, be more comprehensive than traditional databases: Nicola J. Cecchino (2010) found results within Google Scholar that were not present in Web of Science. Conversely, the weakness that has caused a somewhat even playing field between Google Scholar citation profiles and IRs is the lack of authors who have opted into either system. While Google Scholar benefits from already having the metadata of many publications, it too must appeal to scholars in order for them to create the profiles that link these publications. It is likely much easier for a user to sign up for a profile with Google Scholar than it is for them to deposit their works and have a profile created for them in their IR, so it is crucial that we actively promote IRs across our institutions.

Advantages that librarians can share with academics who have or are considering Google Scholar profiles instead of depositing their works in the IR are as follows:

- IRs' clearly defined metrics of assessment in comparison with a Google Scholar citations profile, where Google does not share their methods of calculations and assessment
- Open access to most, and likely all, published works deposited in the IR
- Ability to deposit presentations and other scholarly works not included in Google Scholar results and analysis
- IRs' allowance for adaptation: when Google Scholar is no longer the preferred search method, IRs can adjust to requirements for discoverability on another site

Libraries that are able to succeed in early adoption efforts will have the greatest discoverability within Google Scholar and the greatest chance of growth in the future. Once Google Scholar is able to amass a large number of profiles and if academics continue the trend toward publishing open access works, it will severely limit the likelihood of IRs ever being as robust as they need to be in order to get top spots in search-result lists.

ADVOCATING FOR LIBRARIES

Finally, libraries must continue to advocate for themselves. At one time, there was a push for Google Scholar to mark whether results were open access or not. This tagging would have allowed users to limit their results to open access content, largely increasing the visibility of IRs; however, Google did not choose to make that change. This may change in the future; after all, both Google and Google Image Search did not originally allow you to search by license, but those filters are now available in their advanced searches. Hopefully, in time, that filter will be added to Google Scholar.

In the meantime, mandates like the Office of Science and Technology Policy (OSTP) Increasing Access to the Results of Federally Funded Scientific Research memorandum ensure that open access initiatives will only become more and more important. The SHared Access Research Ecosystem (SHARE) resulted from universities and their libraries laying the groundwork to become more cohesive and structured in their use of IRs as they seek to grant public access to their universities' publicly funded output. This groundwork is based on the presupposition of the importance of IRs in this open access (OA) ecosystem and the importance of standardizing metadata requirements and ensuring those requirements are exposed to search engines and discovery tools (Association of American Universities et al. 2013). Libraries can ensure open access by increasing the robustness and infrastructure of their IRs and helping scholars gain control of their scholarships' dissemination, or they can leave it to others, such as publishers (with their proposed alternative response to the OSTP memo called CHORUS [CHOR, Inc. 2014]), which may not have the same focus on metadata and long-term access that universities and their libraries have.

While Google Scholar may not be completely transparent in the ways in which it indexes IR content, we have enough knowledge to:

* Get content by engaging in outreach to promote IRs for open access
* Focus on unique content (gray literature) or final versions of articles
* Make sure content is findable by optimizing our discoverability

   * ensuring indexing in Google Scholar through the use of metadata
   * shoogling content if necessary

CONCLUSION

In this chapter, we focused on what individual IRs can do to increase their indexing in Google Scholar, but we need to make sure we also emphasize initiatives such as SHARE that allow for the national library community to embrace standardized metadata for the open web and increase the profile of IRs and promote them as integral to the research process. Librarians can and should move toward a future in which we maintain our professional values of transparency and open access by staying aware of the ways in which changing technology impacts our efforts, as well as the shift of scholarly publications toward open access.

REFERENCES

Arlitsch, Kenning, and Patrick S. O'Brien. 2012. "Invisible Institutional Repositories: Addressing the Low Indexing Ratios of IRs in Google Scholar." *Library Hi Tech* 30 (1): 60–81.

Association of American Universities, Association of Public and Land-Grant Universities, and Association of Research Libraries. 2013. *SHared Access Research Ecosystem (SHARE): Development Draft.* June 7. http://www.arl.org/storage/documents/publications/share-proposal-07june13.pdf.

Cecchino, Nicola J. 2010. "Google Scholar." *Journal of the Medical Library Association* 98 (4): 320–21. doi:10.3163/1536-5050.98.4.016.

CHOR, Inc. 2014. "About CHORUS." *CHORUS: Advancing Public Access to Research.* http://www.chorusaccess.org/about/about-chorus/.

Dawson, Alan, and Val Hamilton. 2006. "Optimising Metadata to Make High-Value Content More Accessible to Google Users." *Journal of Documentation* 62 (3): 307–27.

Giustini, Dean, and Maged N. Kamel Boulos. 2013. "Google Scholar Is Not Enough to Be Used Alone for Systematic Reviews." *Online Journal of Public Health Informatics* 5 (2): 214–22. doi:10.5210/ojphi.v5i2.4623.

Google Scholar. 2010. "Inclusion Guidelines for Webmasters." http://scholar.google.com/intl/en/scholar/inclusion.html.

Poynder, Richard. 2014. "Interview with Kathleen Shearer, Executive Director of the Confederation of Open Access Repositories." *Open and Shut?* (blog). http://poynder.blogspot.com/2014/05/interview-with-kathleen-shearer.html.

Part V

Library Productivity

Chapter Twenty-Four

Embedding with Google

Using Google to Optimize Embedded Librarian
Involvement

Samantha Godbey

When Barbara Dewey (2004) coined the term *embedded librarian*, her linguistic inspiration came from the practice of embedding journalists in the military during the Iraq War. Like these journalists who integrated themselves into military units, Dewey argued that academic librarians could gain much from integrating themselves into the many distinct arenas of a university, noting that "embedding requires more direct and purposeful interaction than acting in parallel," consisting of "a more comprehensive integration of one group with another" (2004, 6).

Before there was an accepted term for this practice, librarians were already embedding themselves in many ways. In a broad sense, embedding occurs whenever the librarian moves past waiting for students to come to her, instead making efforts to be where the students are and where learning takes place. Any of these could be considered embedding to some extent: office hours within a department, teaching or co-teaching courses in a department, or creating customized resources, such as a LibGuide.

Although embedding can take place in person or virtually, online and hybrid course offerings provide an opportune environment for embedded librarians. Thirty-three and a half percent of all higher education students were taking at least one online course in 2013 (Allen and Seaman 2014), and many more were enrolled in hybrid and in-person courses with an online component. In this era of online access to students, embedded librarianship often takes the form of embedding in a course via the campus learning management system (LMS). Given the prevalence of learning management systems such as Blackboard, Moodle, and Desire2Learn, the LMS provides a convenient space for pushing customized content and services to a class. Integrating librarians and library resources and services into courses in a sustainable way is more attainable than ever. Further, online access to students and online collaborative tools provide opportunities for librarians to meet students and faculty where they are, without necessarily even leaving one's office.

This chapter provides tips for using the suite of freely available Google Applications to develop a sustainable collaboration with colleagues as an embedded librarian, using Google tools for scheduling, collaborating on documents, and meeting with colleagues and students.

Now a tenure-track librarian at a large research institution, my first experience with embedded librarianship came during an internship as a library student in three undergraduate courses and one graduate course. This experience required collaborating with a supervising librarian as well as the course instructor to provide specialized services. Google tools helped make this virtual internship and virtual embedded experience function smoothly and efficiently.

TOOLS FOR COLLABORATION: GOOGLE CALENDAR, GOOGLE DRIVE, GOOGLE+ HANGOUTS

Successfully embedding in a course requires collaboration and effective communication. Because the librarian is contributing to a course managed by an instructor, this collaboration will at the minimum take place between the librarian and the instructor. It might be necessary to keep others up to date as well, such as a teaching assistant or graduate assistant. Librarians considering embedding must decide on tools to help negotiate this collaboration in order to coordinate events and deadlines, work together on documents, and communicate with one another. Thankfully, Google offers applications that can assist with each of these aspects of a collaboration: Google Calendar, Google Drive, and Google+ Hangouts.

The extent to which tools will be utilized depends in part on whether the librarian and other interested parties are all on the same campus and to what extent they value meeting in person. However, these tools can enhance a librarian-instructor collaboration regardless of location.

Requirements are explained in more detail below, but readers should know that some of these applications require a Gmail or Google Apps account to access, and others offer only limited functionality to users without a Google account.

Google Calendar

One of the most useful Google tools for my embedded internship was Google Calendar. I set up a calendar within Google for the events and actions associated with the courses in which I was embedded and shared this calendar with all course collaborators. The calendar included meetings between myself and the librarian, meetings with the course instructor, and major assignments and events from the course calendar.

After entering standing meetings and course dates into the shared calendar, we then planned backward from each of the assignments and events to identify strategic points at which to push relevant materials and services. For example, three weeks in advance of a research paper, it might be appropriate to share a handout or link to a LibGuide highlighting search strategies and suggested databases. However, if that date coincides too closely with the midterm, we might adjust the timing of that student contact.

Being embedded allows a librarian to provide materials and services to students at their point of need, when they are most likely to use those materials and services. This can be more effective than a single one-shot library instruction course; however, embedded involvement can also be difficult to coordinate because many more distinct actions throughout the semester are required. Having the shared calendar to track these events enabled us to stay aware of what each person was working on and what was coming up next, so that we were coordinating our efforts to connect with students at the most beneficial times. If any changes to the calendar were necessary, the shared calendar was automatically updated for all collaborators.

Although users without a Google account can view calendars, they will need to sign in with a Google account to edit and access all features. For users who do have a Google account, sharing a calendar is very simple. Within your Google Calendar home page, clicking on the name of the calendar in the list of "My calendars" reveals a button for a drop-down menu. Click on the triangle to see options for changing calendar-wide settings, including sharing.

Like most time-management web applications and software, Google Calendar provides the option to set reminders for calendar events. In Google Calendar, these reminders can take the form of e-mail notifications or pop-up notifications on your desktop or mobile device. By customizing preferences within a separate "Embedded Librarian" or course-specific calendar (e.g., "ENG 101 Spring 2014"), you can set preferences that make sense for your embedded commitments in general or for a particular course. This is especially important if you have other calendars in Google. For example, within your embedded librarian calendar, you might designate that all new events will have e-mail reminders at both one week and one day before the event.

These settings would not apply to other calendars, such as a different course calendar or your personal calendar.

Another little-known and potentially useful feature of Google Calendar is the option to receive an automatic e-mail each morning with a daily agenda from a specific calendar. This can be very helpful to ensure that all deadlines are being met, even when interaction with a course is spread out over a semester.

Google Calendar Features

- Ability to set up specialized calendars for a particular course or sections of a course.
- Ability to share calendars with collaborators.
- Syncs across devices.
- Works seamlessly with many calendar applications for mobile devices and tablets. Can be viewed alongside Lotus Notes calendars in the iPad Mail application, for example.
- Receive e-mail reminders and/or pop-up notifications for calendar events.

  - Customize default reminder settings within a particular calendar within Calendar Settings.
  - Add reminders to individual events as needed.

- Customize colors to coordinate with other calendars.
- Individual events can easily be copied into other calendars.
- Ability to easily toggle visibility of the calendar, for example, if you don't want a calendar to be visible on your mobile phone.
- Ability to adjust settings for privacy. The default is private, but calendars can easily be made public within Calendar Settings.
- Option to receive a daily agenda for a specific calendar via e-mail.

Google Drive

While time management is extremely important, particularly when coordinating with others, the ability to collaborate on the creation of materials is equally important. Over the course of a semester or multiple semesters as an embedded librarian, the librarian will create numerous documents in support of the courses in which she is embedded, whether these are planning documents to be shared among collaborators or documents that will be shared with students.

Google Drive, formerly known as Google Docs, is a Google service that enables users to create, store, and access files on any of their devices. Options for accessing Google Drive include via any web browser, on the hard drive of your computer, or with mobile apps. The suite of productivity applications in Google Drive parallels the kinds of programs many users are familiar with already—Document (similar to Microsoft Word), Presentation (similar to PowerPoint), Spreadsheet (similar to Microsoft Excel), Form (a survey tool), and Drawing. Users can create word documents, presentations, spreadsheets, surveys, or drawings and allow others to view and edit those documents.

In my internship, documents, spreadsheets, and forms were all put to use as an embedded librarian. A shared document was used to manage an ongoing agenda in order to make transparent who was working on what at any given time. This shared agenda formed the basis for each weekly meeting between myself and the librarian and helped document accomplishments throughout the semester. During each meeting, especially videoconferences, both contributors were able to view and edit the document. Shared documents were also used to draft announcements that would be submitted to students in the LMS and to collaboratively develop other resources for students.

Working on a shared document in Google Drive allows multiple collaborators to contribute to a document as it is being developed without being physically together. One method used by collaborators is to e-mail revised versions of a document back and forth. However, at any given time, one user is denied access to the document. The same is true for those users who have the option of placing collaborative documents in a shared drive on their institution's intranet, as documents on shared drives are generally restricted to a single user at a time. If one person is using a document, a second user will be unable to view new edits or save any changes until the other user has closed the document.

Using Google Drive eliminates the cumbersome process of exchanging versions of a document or waiting for one another to finish edits. With Google Drive, multiple users can access the same document at the same time. Editors can see one another's edits in real time. Each user is given a different color cursor, and the name of the editor appears alongside the cursor so that each person knows who is making changes.

In addition to simultaneous editing, Drive allows users to comment on their shared work. Comments appear in a commenting pane to the right of the document, allowing the exchange of ideas without cluttering up the document itself. Comments are threaded as discussions, so individuals can respond to comments made by other users and mark a comment thread as "resolved" if appropriate.

Finally, although not directly related to its benefits for collaboration, one feature many users value is that Google Drive automatically saves all documents every few seconds. Users need not worry about lost work due to power outages or other disruptions. At the same time, Google retains copies of earlier versions of each document, and it is simple to revert to an older saved copy. This capability might put some collaborators at ease in case they decide to return to an older instance of the document.

As with all documents stored on the Internet, privacy is an important consideration when working with Google Drive. Similar to Google Calendar, the default sharing settings for Google Drive make a document private, so that only the creator can access that document. Under the File menu, select "Share" to enable others to collaborate on the document. Documents can be shared with specific people, with anyone to whom you provide the link, or with anyone on the Internet. The "Public on the web" option enables a document to be found using a search engine. This may be appropriate for some documents; however, it is likely that for most works in progress you will want to select one of the stricter privacy settings.

Another option for sharing a final version of a document with students is to publish a copy of a Google Drive document to the web, at which point a unique URL is created for that published copy. Visitors to the published document will not be able to edit the document, which is a separate copy from the original file. The original document retains its privacy settings regardless of whether it has been published.

Google Drive Features

- Upload or create shared documents and store them online. E-mail links or download documents to various file formats, including .doc, .ppt, and .xls.
- Ability to access documents on any device via http://drive.google.com.
- Option to install Google Drive on your computer or Google Drive app for your smartphone or tablet.
- Multiple users are allowed to edit documents at the same time.
- Syncs across devices.
- Automatically saves documents.
- Automatically saves revision history, including who made edits.
- Ability to revert to older versions of a shared document.
- Documents can be organized into folders. Entire folders can be shared with a group of collaborators.
- Currently 15 GB of storage is allowed across Google Applications. Additional storage is available for a monthly fee.

Google+ Hangouts

While Google Calendar and Drive provide the means to coordinate work with colleagues, and Google Drive enables users to interact in the same virtual space within the confines of a specific document, Google+ Hangouts is the most comprehensive Google tool for real-time communication. Often referred to simply as Hangouts, this instant messaging and video chat platform allows users to communicate with one another directly. Depending on one's situation as an embedded librarian, it might make sense to have face-to-face meetings, but Google+ Hangouts provides the means to interact instantaneously, regardless of location. The emphasis in Hangouts is on sharing things in real time—whether that is text, voice and/or video, an image, a YouTube video, a document, or your own desktop. Hangouts provides the means to have a shared experience—online.

Google+ Hangouts can be accessed through the contacts list in Gmail, via Google+, or through an app on a mobile device. The instant messaging client works in the same way as other instant messaging and chat programs, with users typing messages to one another and hitting return to send. Like other programs, Hangouts indicates when the other person is typing, helping to prevent interrupting one another or sending messages at the same time. Unless the users are also enrolled in Google+, Google's social network, this is the extent of the instant messaging capability. If users are registered as Google+ users, they are able to embed photographs in their messages and participate in group chats with up to one hundred participants. Much like with instant messaging, voice and video calls are also limited to one-on-one conversations for non-Google+ users. For Google+ members, voice and video calls can include up to ten people.

Any contact can be invited to a Hangout. However, a contact who does not have a Gmail or Google Apps account will be sent an invitation to join the service in order to chat with you.

Even when in-person meetings are an option, having a video-conferencing application such as Hangouts provides the flexibility to meet as needed. My internship supervisor and I met regularly via video chat to accommodate our differing on-campus schedules. This allowed us to meet consistently, regardless of our locations.

Another benefit of using the Hangouts instant messaging service for quick exchanges that arise while embedded is that Hangouts maintains an ongoing record of messages. This makes it easy to resume a conversation or go back to the archive to see exactly what comments were made and when. Because it is integrated with Gmail, e-mail and messaging history are all in the same location.

Google+ Hangouts Features

- Options for messaging, video, and voice. Ability to have free voice and video conversations on your computer.
- One-on-one or groups. Up to ten people with video, up to one hundred people without video.

- Integrated into Gmail. Users can initiate a text, voice, or video conversation directly from Gmail.
- Within a Hangout with Google+ users, includes options to share your screen, open YouTube or Google Drive, and use other helpful apps to be shared within the Hangout.
- Maintains record of messaging history.
- Ability to start a Hangout and switch from one device to another if needed, for example, from a desktop computer to a mobile phone.
- Users who have joined Google+ can easily initiate a Hangout with a specific Google+ circle.
- Both users must have a Gmail or Google Apps account.
- Google+ account required for group video calls and sharing photos.

TOOLS FOR CONNECTING WITH STUDENTS: CALENDAR, GOOGLE+ HANGOUTS, GOOGLE FORMS

In addition to the collaboration and communication between the librarian, instructors, and others involved in teaching the course, individuals considering embedding in a course or hoping to improve their embedded experiences must remember that the main goal of embedding in a course is to enhance student learning. Several of the tools already discussed can be instrumental in connecting with students. More tips for these tools are included here with an emphasis on how they can be used for communication with students.

Google Calendar

Users should remember that calendars can be used to coordinate the schedules of the librarian and collaborators and also to schedule contact with students. As discussed previously, in my experience embedded in multiple courses, a shared calendar was used to plot out strategic actions within a course, and reminders helped ensure follow-through on action items. If we decided to share a particular handout or link three weeks in advance of a research paper, a reminder would be set to coincide with that action item. Likewise, if an assignment required a particular database, a reminder would be set to share database search tips in the day or two following the distribution of the assignment. A reminder would be set to contact students a week before a large project was due, encouraging them to set up research consultations. Google calendars can also easily be embedded in web pages to share with students.

Google+ Hangouts

In addition to using Google+ Hangouts to communicate with colleagues, Hangouts can be an excellent resource for interacting with students. Particularly if a librarian is embedded in an online course or a hybrid course with limited class time, she may have limited face-to-face contact with students even when she is a persistent online presence in the course. Hangouts can also be used for meetings with individual students or groups. As an embedded librarian who is trying to meet students where they are, the option to meet virtually can be indispensable, particularly for distance students.

If students are Google+ users, they will be able to join in a group meeting or help session with up to ten students. The librarian can choose to share her screen, allowing students to view a demonstration in real time. Alternatively, a student can share his screen to allow the librarian to view the student's search process or help troubleshoot technical problems.

Hangouts Features for Use with Students

- Increases flexibility of schedule for meeting with students—for students who are not able to make it to the librarian's regular hours.
- Meet with and provide assistance to groups of students at the same time—up to ten with video, up to one hundred with text only.
- Screen-sharing feature enables the librarian or student to see what the other is doing.

Google Forms

One of the applications in the Google Drive suite that can greatly assist with meaningful student interactions is Google Forms. Google Forms allows users to create surveys and quizzes and publish them online. Responses are visible to the form's creator within Google Forms or can be sent directly to Google Spreadsheets.

Creating a form is a straightforward process, starting with the selection of a design theme and adding questions. There are a number of options for types of questions, as well as the option to create multiple pages of questions in order to send a respondent to a certain page based on a particular answer.

Forms can be used for feedback or assessment purposes. We used Google Forms, for example, to gather feedback from students on our involvement in their courses, asking questions about what they had found to be most helpful during the semester. We posted the link to the student survey to the course site in the LMS and customized the confirmation page to include a custom message thanking participants and directing them back to the course Lib-Guide.

Aside from ease of use, a benefit of Google Forms is that, as with the other Google Drive applications, multiple collaborators can view and edit the form and results. I recommend choosing a Google spreadsheet as the destination for responses. The responses can then easily be sorted within Google Drive or exported and manipulated within Microsoft Excel or another program. Entries include the time the responses were submitted, so the same form can be used for multiple semesters without confusion.

Regarding sending the form to students, it is important to note the difference between "sharing" the form and inviting "collaborators." Collaborators are able to edit the form. Be sure to provide the link under "sharing" to students.

Google Forms Features

- Many themes to choose from, currently nineteen, ranging from simple to more colorful.
- Ability to set different types of questions based on the answer format: text, paragraph text, multiple choice, checkboxes, choose from a list (a drop-down menu), scale, grid, date, or time.
- Ability to insert images or videos.
- Ability to send responses to a shared Google spreadsheet.
- Ability to disable a form and reopen at a later date if desired.

GOOGLE TOOLS FOR SUSTAINABILITY

For many librarians, the question of whether to pursue the embedded model is quickly dismissed given their already substantial workloads. However, each of the tools discussed in this chapter can help to establish an effective and sustainable collaboration with course instructors. Google Calendars can help to alleviate the potential anxiety of managing ongoing deadlines by setting customized alerts for action items. Google Drive can help to create truly collaborative documents that are stored in an accessible place. Google Forms allows librarians to collect data from students and compare it across semesters. There is, of course, a time investment as the librarian orients herself to new tools and establishes systems that work for her, but ultimately, these tools allow one to share work with others, automate components of the work, and acquire a portfolio of materials that can easily be reused from semester to semester.

CONCLUSION

As with all Google products, these projects are works in progress, and these exact numbers may change over time. Google Hangouts is a relatively new product, having integrated features of the now defunct Google Talk, and speculation abounds that soon Google Voice will be incorporated into Hangouts as well. These types of products that allow users to share calendars, store and collaborate on documents, and communicate synchronously and asynchronously will continue to exist in some form. An awareness of how these products can help a person manage, implement, and document one's work will allow a librarian to do so effectively, regardless of what the newest version of the products are called. These tools from Google can benefit any embedded librarian deep in the academic trenches.

REFERENCES

Allen, I. Elaine, and Jeff Seaman. 2014. *Grade Change: Tracking Online Education in the United States*. Babson Survey Research Group and Quahog Research Group. http://www.onlinelearningsurvey.com/reports/gradechange.pdf.

Dewey, Barbara I. 2004. "The Embedded Librarian: Strategic Campus Collaborations." *Resource Sharing & Information Networks* 17 (1–2): 5–17.

Chapter Twenty-Five

Librarians' Everyday Use of Google Tasks, Voice, Hangouts and Chat, Translate, and Drive

Fantasia A. Thorne-Ortiz and Neyda V. Gilman

Students rely heavily on Google for research, e-mail, and everyday web browsing. Librarians use Google for the same reasons (yes, even for research!), but our use of its productivity tools and software continues to increase as Google's capabilities and resources expand. This chapter will highlight a number of tools, including Google Tasks, Voice, Hangouts and Chat, Translate, and Drive, documenting their development, current library use, and ideas for future use by librarians.

GOOGLE TASKS

Librarians have a lot going on at work and at home. Google Tasks is a simple, streamlined tool that makes it easy to organize life and remember tasks. Tasks appears in the Gmail window, similar to a chat box, allowing you to have your list always visible without it getting in the way of e-mail. It can easily be minimized, popped out, or opened full page in a separate tab. Always on the go? Tasks is easy to use on any mobile device with a compatible browser or by using a third-party app such as GTasks or Tasks.

Development

In late 2008, Google Tasks was introduced to Gmail Labs, Gmail's testing ground for new products. It was a lightweight to-do list with simple tasks that could be checked off as completed. A few months later Google took Tasks mobile, allowing lists to be managed from Apple or Android mobile devices. In May of 2009, Google Calendar was linked with Tasks, allowing tasks with due dates to appear in Calendar. Tasks spent less than a year in Labs before it became a feature on all Gmail accounts. Additional improvements include the ability to:

- easily move tasks between lists,
- merge and split tasks,
- print the tasks lists, and
- integrate with the Thunderbird mail client.

For more information on the history and growth of Google Tasks, visit the *Official Gmail Blog*.

Like many other Google products, Tasks was not made to be complex, but there is more to it than its simple design suggests. You can create multiple lists, have categories with subtasks, make notes, easily turn e-mails into tasks, assign due dates, and link with Google Calendar, Thunderbird, and Outlook.

Current Library Use

So what can you do with Google Tasks? Let's look at an example. Evelyn, "Evy," is an academic librarian with two cats and a dog. Examples of Evy's Tasks lists include work, home, and groceries.

Here is how she uses them.

Her groceries list is just a simple list of items. She can add to the list whenever she thinks of something she needs, either through her computer or her phone. If it is something for a recipe, she just adds a quick note to make sure she gets the appropriate amount. While out shopping, she uses her phone to check off items. There are many apps that are designed for the singular purpose of managing grocery lists, but Tasks provides this in addition to a plethora of other services.

Her work and home lists get a little more complicated. Under her home list, she has main tasks for each cat, Ginger and Snap, and her dog, Brutus. She then makes tasks for appointments and other events as they come up. Ginger has a subtask reminding Evy to take her to the vet on Tuesday, Snap needs to get her claws trimmed, and Brutus has a couple of playdates next week. Ginger's vet appointment has a due date that syncs with Evy's calendar. This way she can see it in her calendar even if her Tasks list is closed. She also set up reminders to make sure she doesn't forget. Brutus's playdates work the same way.

The work list also uses the subtasks feature. Evy has main tasks for instruction sessions (which she labels classes), professional development, and meetings. As with her home list, additional tasks are added as necessary. Her classes task has subtasks for all of the instruction sessions she has scheduled so far this semester. Each session then has subtasks depending on what needs to be done. A feature Evy really likes is being able to make notes for each of the sessions to help her remember what the different professors have in mind or to jot down quick thoughts about the class. When Evy is co-teaching with another librarian, she e-mails the session task list to her co-teacher. The other instructor can't edit Evy's tasks, but can see what needs to be done. This can be helpful for any group project.

Her professional development list is mostly made up of e-mails about webinars, conferences, or articles she wants to read. When she gets an e-mail she wants to save, she simply clicks "Add to Tasks." This will put the e-mail in her tasks list where she can then move it to the appropriate list, assign a due date, and make notes.

Under meetings, there are subtasks for each meeting, and some meetings have additional subtasks so that she shows up prepared. All of the meetings have a due date to ensure they show up in Calendar. Evy will sometimes make a separate list for a meeting where there are multiple points she wants to talk about. She can then print the list and take it with her to the meeting.

One problem that Evy has run into is that the institution where she works doesn't use Gmail. They use Outlook. Since Evy loves using Google Tasks and doesn't want to worry about updating multiple lists, she pays for the third-party software CompanionLink to sync her Google Tasks with her Outlook. Evy also uses Chrome and installed the Chrome Tasks extension. This allows her to get to her tasks list with a click of a button.

The Together Teacher blog recently suggested another way to use Google Tasks. They suggest making a task for each month and then making subtasks for everything that needs to be done in that month. Everyone has their own way of using Google Tasks. Play around and find what works best for you.

Ideas for Future Library Use

It is hard to say what the future holds for Google Tasks. There are many other electronic to-do lists available. Some people think that the new Google Keep (part of Drive) will be the death of Tasks. Tasks is simple and streamlined. It has also been out long enough that there are third-party programs that make using Tasks more versatile. Google provides an application programming interface (API) that makes it easy for developers to continue building programs that integrate Tasks, making it even more useful to a wider range of individuals. If you have a Gmail account, give Tasks a try. In the top-left corner of your e-mail click the down arrow next to Gmail and select "Tasks." That is all there is to it. Go get organized!

GOOGLE COMMUNICATIONS

Google has worked hard to live up to its goal to make information, including communication, accessible and useful. Beyond Gmail, Google has developed and improved multiple programs that facilitate communication. This section will cover the various communication programs, including Talk, Chat, Voice, and Hangout.

Development

Google Talk, introduced in 2005, was launched to enable users to make phone calls and send text messages via the Internet. Six months later, at the request of users, Google added a chat service. Initially the calls, texts, or chats were limited to those between Google users. This changed by the end of 2007, when Chat expanded to integrate other services such as AOL Instant Messenger.

Behind the scenes, another telecommunications service was being adapted from a competing program called GrandCentral. In 2010, the transformation from GrandCentral was complete, and Google Voice was released to the public. Voice users are assigned a phone number and have the option to connect all of their phones together so that calling one number rings all phones. Having one number to rule them all allows calls to be made over the Internet to other Internet users, as well as to landlines and cell phones.

Google is always trying new products, including their communication products. There have been various versions that were either created from scratch or adapted from acquired products. It can be confusing trying to figure out if you are using Talk or Chat or Voice or even e-mail (both Chat and Voice voicemails can be sent and stored in your Gmail). Depending on exactly what you're doing and what device you are using, the name of the product can vary. The endgame of all of this is Google Hangouts. Originally released as part of Google+, Hangouts has branched out onto its own. It allows for phone calls, video calls, chats, and texting. Hangouts also allows for groups of people to participate in any of its services at once, so group calls and discussions are easy to make happen.

Current Library Use

Chat

There are many ways to use these Google services, both in our personal and professional lives. Something as simple as Chat can make the day run a little smoother. Running late for a reference shift or wondering where your re-placement is? Have a quick question for another librarian? E-mail inboxes can easily become inundated with one-line e-mails. Sending an instant message (IM) is a quick and easy way to say what you need to say. A Chat session pops up in your Gmail and is a good way to have a quick conversation without overwhelming inboxes.

Voice and Hangouts

Most people use at least two phones every day. There are work phones, cell phones, and home phones. Some people may even have a Google or Skype number. Voice allows all of your phones to ring whenever someone calls any of the numbers. You choose which phone to answer. If you don't answer and they leave a message, that message can be e-mailed and/or texted wherever you'd like. If you would rather not answer the call from your doctor at work, let it go to voicemail and then read the transcribed voicemail in your e-mail.

This same thing can be done at the library by connecting the reference phone with Voice. Calls will go to the phone (and the associated e-mail if so desired), and text messages will arrive through e-mail and any linked cell phones. According to a 2012 Pew Internet report, teens are texting more and e-mailing and chatting less (Lenhart 2012). Using Voice or Hangouts makes it possible for a librarian to easily communicate with a diverse patron base. Whether it is a phone call, voiceover IP call, video call, text, IM, or e-mail, Voice/Hangouts is all the librarian needs to be there for the user.

Future Library Use

Hangouts has been used for a variety of events, including a question/answer session with Sherman Alexie during Banned Books Week, by NASA, and even by President Obama. It is currently being used for library professional development, mostly using the Hangout On Air service. The American Library Association (ALA) has a series called American Libraries Live where a panel of librarians discuss a topic. Each of the librarians logs on from their office to join the video call. This "call" is streamed live, so anyone can log in and watch the discussion. Viewers can contribute to the discussion by participating in the chat and/or using Twitter with the appropriate hashtags.

What else can Hangouts be used for? There are many possibilities. Ideas include:

- classes, especially with distance learners;
- another reference service point, beyond chat;
- virtual conferences; and
- professional development sessions similar to American Libraries Live, rather than a traditional webinar.

A concern regarding Hangouts and video chats being used during reference transactions is the issue of privacy. Anyone should feel comfortable asking a librarian for help and video can make people uncomfortable. However, video chats can be beneficial for those who work better talking face to face but cannot make it into the library. We all know the power of a properly conducted reference interview and how difficult that can be when not meeting in person. Video calls can help the librarian build a relationship with the patron and work through the problem together. This may prove to be especially useful for distance learners or those with in-depth research needs who can't make it into the library. Having a variety of communications methods allows users to get the help they need in a way in which they are comfortable.

Any time Hangouts is used the transaction can be kept private, streamed live, recorded and shared on YouTube, or both streamed and recorded. It is even possible to share screens, making it easy to demonstrate a search or see what the other person is talking about. There are lots of ways to communicate with Google, so get the app or log in to your account on a computer and start exploring!

GOOGLE TRANSLATE

There are a number of translation services and tools offered online, such as Yahoo BabelFish and Bing Translator. Google Translate is a leader in the translation arena, and this free service currently supports over seventy languages, including Arabic, Greek, Hebrew, Spanish, and Thai. Google Translate can translate a word, phrase, and also an entire web page. Although the translations are not always precise, the service will provide immediate assistance to those with access to computers or mobile devices. Google Translate's development, current use in libraries, and implications for future use will be discussed in further detail below.

Development

A system called "statistical machine translation" creates the translations provided by Google Translate, and this method has been in use since 2006. Millions of documents are entered in the machine translation system, creating patterns that produce the finished translation. As more documents are entered into the system, translations are produced faster and more accurately. Some languages are not as precise as others due to the number of documents the translations are accessing, but as more translated documents are provided, translation quality will improve. To access the list of the languages Google Translate supports, visit "Inside Google Translate" online (http://translate. google.com/about/intl/en_ALL/).

According to an April 2012 Google blog post, more than two hundred million people use Translate every month, and those numbers are sure to have increased. The amounts of translations that occur on a daily basis were likened to the amount of text found in one million books. It is obvious Translate has come a long way since its initial beginnings in 2001, when it took a little over two days and a thousand machines to translate a thousand sentences. Currently, using Translate on a mobile device opens a new world of capabilities. There is also the option in Gmail to translate e-mails. Google has been diligent in its support and development of Translate, and there are a number of ways librarians can make use of its translation services.

Current Library Use

Have you ever found yourself in the position of trying to assist a patron, but there is a language barrier and no apparent way to overcome it? Has it ever been assumed you speak a language that you don't? On a number of occasions I have found myself trying to speak a language I studied years ago and feeling confused and a bit awkward because I'm not sure I'm communicating effectively. Language barriers can occur in airports, stores, the community, and in the workplace. Google Translate may be the answer to dilemmas such as these, because it is available on any device with an Internet connection and is downloadable on most mobile devices. We may not carry mobile devices around with us at all times, but it may be a good habit to practice when we are in public spaces in the library.

If you aren't comfortable trying to speak another language, consider sitting down with a patron at a computer and typing out the conversation using Google Translate. After you choose the languages you are translating from and to, enter the text as a sentence and you can watch it translate in the second box. There is a feature in the first text box that allows you to speak the words you want translated if you have a microphone attached to your computer. The second box also contains a microphone icon, as well as the option to listen to the text that has been translated. This feature is especially helpful, and turning on your speakers will allow the patron to hear the translated text.

Ideas for Future Library Use

As Translate's abilities expand, librarians should continuously consider methods of using the tool to assist patrons. The roadblocks forced upon us by language barriers are breaking down, and staying current on new developments in technology and software will ensure our success in providing resources to our patrons in the future. Google Translate's mobile app allows users to use speech or handwriting recognition when entering text for translation, and these tools can be used during library instruction sessions. English as a Second Language (ESL) classes often come into the library for instruction, and using Google Translate on tablets or computers could assist the instructors as well as the students with communication barriers.

Youth librarians can make use of Translate for story times and potentially broaden the populations they reach. If a librarian has a bilingual book they are considering using during story time, but is unsure how to pronounce some words, Translate is a useful tool for assistance with phonetics. Academic librarians interact with diverse student populations and also communicate with community patrons on a daily basis. For example, if there isn't a staff member who speaks Spanish in your library and a patron isn't fluent in English, grabbing a smartphone, tablet, or quickly going to a computer to use Google Translate, is a better option than turning the patron away. Devise a plan of action for staff in advance of interacting with a patron who is not fluent in English to avoid scrambling when the patron is in need of immediate service.

Translate now provides translations for text in photos and YouTube video captions, so using the service will provide librarians with the capability to assist patrons with almost every type of media that exists. The days of being unable to translate words, sentences, and web pages are over. While human translations are ideal, we currently have more options than trying to find a translator if we encounter a language hurdle either during a conversation or while reading a document. Do you want to say "Hello!" to a Spanish speaker and you don't know how? The word *¡Hola!* can be spoken by you or directly from Google Translate to the patron faster than you can imagine. Get the app, go online, and give it a try.

GOOGLE DRIVE

Many librarians use a number of different computers on any given day. We often e-mail a document back and forth to ourselves or others, and it can be confusing as to which document is the most up-to-date version. Do you remember saving a document using flash drives? Floppy disks are a distant memory now, and the ability to access documents in that fashion is almost extinct. The capability to save and create documents is becoming increasingly streamlined, and even flash drives are serving as "backup tools" instead of the main repository for our documents and files. Cloud storage of books, images, movies, and documents is currently the norm, and Google Drive continues to make advances in providing a one-stop creation, collaboration, and storage service. Google Drive's possibilities are endless, and using Drive to its fullest capacities will provide librarians an edge when it comes to collaboration.

Current Library Use

Drive is increasingly becoming the go-to web service and application that professionals use to collaborate. Google Drive was created in 2012, and is currently the home of document-creation tools, including Google Docs. With the capability to store books, photos, and various types of documents, Drive is becoming a part of our everyday lives.

Some may prefer and rely on Microsoft services such as PowerPoint, Word, and Excel, but when there is a need to collaborate or work remotely, the initial document creation process can begin in Google Drive. Upload any Word document and a group of people can contribute to the same document simultaneously. A common concern with Drive is that public documents are searchable on the web. Choosing to have your documents remain private, or only viewable to those you have shared them with, can assuage some concerns over privacy. Google Docs was expanded into Drive as competition to DropBox, the familiar file-saving system, as well as other cloud-storage companies.

The article "Cloudy with a Chance of Collaboration: How Cloud Services Can Enhance Collaboration among Librarians," by Carla Wale and Ellen Richardson, summarizes a few resources such as DropBox, SkyDrive, iCloud, and Google Drive that allow users to save data using cloud storage. Quotes from librarians are also included, indicating the cloud-storage tool they prefer to use. The list below provides examples of how librarians have made use of Google Drive to date:

• Use the document-creation tools for projects, spreadsheets, presentations, and more
• Project collaboration with local, regional, or national colleagues
• Create notes and meeting minutes
• In place of Microsoft software for personal use
• Use during instruction sessions for student contributions

Continued Advancements in Drive

Keeping abreast of Google Drive's continuous updates will allow librarians to stay on the cutting edge of software and technology. For up-to-date information, visit Google Drive's blog, where you can read about updates such as handwriting input, real-time text cursor in Slides, and more. Videos and images can be uploaded to Drive, and while they must be downloaded to your computer to view, it is becoming an increasingly streamlined tool for storage of all types of media. Search "Get started with Google Drive" online to read instructional details about using Drive.

Some librarians may have been introduced to Drive in library science school or through word of mouth by other professionals. Ideally, librarians will continue to use Drive to its fullest capabilities while continuing to create new methods of collaboration with the tool. With 15 GB of free storage the possibilities are endless, so let the creation, collaboration, and storage begin!

CONCLUSION

Google is continuously creating and updating cutting-edge technology and software. We can't predict the future or Google's next steps, but we believe the capability to communicate, collaborate, and stay organized using their free resources will survive the test of time. A best practice for librarians is to keep current of Google's developments in order to maintain productivity and reach our patrons with services they are already using. Delve into these tools to fully explore their capabilities and their possibilities in your work and everyday life.

REFERENCES

Lenhart, Amanda. 2012. "Teens, Smartphones & Texting." *Pew Internet & American Life Project*.

Wale, Carla, and Ellen Richardson. 2013. "Cloudy with a Chance of Collaboration: How Cloud Services Can Enhance Collaboration among Librarians." *AALL Spectrum* 17 (6): 10–12.

Chapter Twenty-Six

Google Alerts, Trends, and Chrome in Public Relations

Chelsea Dodd

The effective practice of public relations (PR) requires constant awareness of what is going on within the library and around the community. It is important to identify the library's brand—that is, what distinguishes your organization from others in town. Whether it is a logo or color scheme, your brand is how you want your community to view your institution.

If your library is like mine, there is not room in the budget to hire some-one for public relations. Furthermore, practicing good public relations takes time—another commodity limited for most library staff. While still new to my position at Montclair Public Library and getting reacquainted with New Jersey, I began doing regular virtual public relations exercises to get an understanding of the library's position in town. Using Google resources, I have found that the routine processes of public relations can be handled methodically and with limited staff hours. The most useful tools include Google Alerts, Google Trends, and Google Chrome. Each of these are quick to set up and easy to use—and that's something professionals in any industry can appreciate.

GOOGLE ALERTS

An important component of savvy public relations is knowing what is al-ready out there. While technology does save time with certain functions, the size of the virtual world makes it almost impossible to monitor everything on the web. This is where Google Alerts comes into play.

Google Alerts allows for scheduled e-mail notifications on predetermined search queries. So, for instance, you can set up an alert to be sent whenever the library is mentioned in a news article. This e-mail will include a link to the article and, depending on the content, you can opt to publicize the article within the library's networks by cross-posting the link. By encouraging traffic to the web page where the article lives, this common activity maintains the mutually beneficial attribute of true public relations.

Found at google.com/alerts, this search feature is intuitive to manage. The best approach to cover all bases when doing general searches on a public library is to set two queries, both in quotation marks: "[Town] Public Library" and "[Town] Library." For academic or special libraries, just the name in quotation marks should suffice. Keep in mind that you may get irrelevant results with this general approach, but pertinent information will not be missed.

It is also possible to use the minus sign (-) operator or the "site:" operator to exclude known conflicts. For instance, there is a Montclair Public Library in New Jersey and a Montclair Branch of the Oak Lawn Public Library in California. Searching for "Montclair Library" brings up results including both libraries. By adding "-California" to the search query, results including mentions of the California library are omitted.

The next step is to decide on the result type. These types are based on the Google Search categories that most library staff are already familiar with: everything (i.e., surface web), news, blogs, videos, discussion, and books. While "news" is the preferred category for local articles, it is important to know *everything* that is out there.

The language and region settings are self-explanatory. Depending on the community, the language could be a factor. For instance, if there is a large Spanish-speaking population, it may be a good idea to set up search parameters using Spanish search terms. With regard to the region, it is rarely necessary and, since the globe is getting smaller thanks to the expanding virtual world, it is best not to limit by location.

The other two options to decide are alert frequency and best results or every result. These are strictly by preference. Weekly alerts tend to be the more economical with time. It is easier to scour a few weekly e-mails one morning a week than to be distracted by e-mails at various times throughout the day. Only alerts that have time-sensitive content should be scheduled "as it happens." As mentioned earlier, this includes news articles that could be immediately posted to the library website and shared via social media.

As you gather these alerts and look at the content, pay close attention to the perceived attitudes and feelings that the community has about the library. Are they negative news articles? Do the blog posts have positive commentary? These alerts bring the library awareness of its current brand and can be used as evidence supporting a public relations campaign for a brand overhaul. Export relevant results under the "Manage your alerts" web page, and share with colleagues, trustees, and anyone else with sway over the direction that the institution takes.

Once alerts are set for library-related items, it is also practical to set up alerts for queries involving the town and other local organizations. This will help you get a feel for hot topics that your library can address through programming. It will also show what the media and those with an online presence are interested in circulating with their readership. There is no need to reinvent the wheel, especially when it comes to public relations pitches that are working. It is prudent to reflect on others' strategies and contemplate ways that they can work for the library. That being said, exact replicas of pitches will be ignored.

While libraries may not be competing with other organizations in the traditional sense, every entity in a community is vying to attract those they serve. Creating queries for the local YMCA, museums, student clubs, and other cultural and civic institutions will help in evaluating town interests. Try several variations and do not be afraid to shake things up every few months.

GOOGLE TRENDS

Another great way to monitor the virtual world is using Google Trends. Found at google.com/trends, this search feature "analyzes a percentage of Google web searches to determine how many searches have been done for the terms you've entered compared to the total number of Google searches done during that time" (http://support.google.com/trends). Like Google Alerts, Google Trends also has a subscription option to get e-mail alerts of set search terms once a week or once a month.

By exploring what is trending in the information sciences world and in the geographic vicinity, you can discover new ways to support your community as well as ideas for pitching library services. Being in New Jersey, I created one basic search for "library" with the location "New Jersey." What is fascinating but not surprising to see is the sharp rise in searches for "library" from 2008 to 2010. The decline has begun again now that the recession is fading from view. The goal identified by this search is halting the decline, and the challenge is finding ways to be relevant as the economy strengthens.

When you look at the rising "Queries" under Related Searches at the bottom of the screen, Parsippany Library has a "Breakout" percentage. This means that the percentage of people in this region searching for "Parsippany Library" has grown by over 5,000 percent since 2004. By exporting this list, I can see further results, including the disheartening number of searches for "Montclair Public Library."

Why should I care? Recall that efficient public relations involves not reinventing the wheel. By seeing what people are searching for and what libraries are coming out on top, I can perhaps pick up a few pointers by examining those libraries' websites, social media accounts, and programming. Maybe that thriving library had a plan, or maybe their trending success is purely incidental. Lucky for us, the library industry is full of professionals willing to share their tales of success and failure. Do not be afraid to pick up the phone and speak with someone from a library with clearly flourishing communication achievements.

Trend searches should also include general terms focused on new library movements and localized issues, for instance, *makerspace, job growth*, or *sustainability*. These results can be analyzed to get a feel for where these issues are popular—Does your state lack these interests? Are most results clustered around your town?—and that analysis can help you determine if there is room for growth in promoting such educational programs and services.

Public relations is never a simple activity, but by gathering knowledge about trends, libraries can better equip their public relations toolkit. It helps save staff hours by not wasting time on interests that are in sharp decline or projects that are not unique. There is no point in pitching a story to a reporter or editor if it has little chance of getting noticed. Save your energy for the ideas that are trendy and valuable to the media that would run it. The hardest thing for most people is recognizing a lack of a story despite their elation for the program or initiative. When in doubt, do not be afraid to ask people outside the building and library world for their input.

GOOGLE CHROME

Google Chrome's browser appears to be slow taking off in the library realm. Its sleek design and features outweigh those of its competitors, however, and library staff looking to integrate Google products should consider making the switch. The following are some of the top reasons why you should download it today at http://www.google.com/chrome.

Chrome Profiles

To start, Chrome allows the creation of profiles that can be used to access specific settings from any computer. Developing one profile with PR-focused configurations enables those active in a library's communication plans to share saved usernames and passwords, tabs, Chrome extensions, and more. This profile ensures that all accounts and information are accessible by team-working staff. Creating a collective public relations Google account to take advantage of sharing a calendar of events and deadlines and Google Drive documents such as media lists develops synergy among collaborators.

For those smaller libraries with only one person dedicated to handling public relations and social media, this profile can still come in handy. Separating the communication functions of the Chrome browser from personal accounts can be more productive from a workflow standpoint. It is simpler to switch profiles to focus on tasks to be accomplished than it is to navigate between different Google account logins from the same window.

If your institution has a high staff turnover rate, a Chrome profile also helps create a foundation for the library's communication activities that can be continued by future employees. As someone who has had to start from scratch in nearly every new job, I try to be cognizant of leaving thorough information written down before I go. Creating a profile ensures that the public relations life of the library can be easily located and sustained after the current staff departs.

Chrome Extensions

Similar to Internet Explorer or Firefox add-ons, Chrome extensions allow users to install more browser functions via buttons. These extensions make saving and sharing information quick and easy. With the plethora of developers adding extensions to the Chrome Web Store, the ones discussed in this chapter are those that I found the most effective and, more important, free. By saving these extensions in the aforementioned public relations Chrome profile, there is no need to configure each participating staff member's browser. Just add the profile and sync the data under the browser's settings.

Any.do

The value of writing to-do lists is no secret. Most library staff have even toyed with using bulletin boards chock-full of tasks or dry-erase boards filled with deadlines. The Any.do extension takes these ideas into the twenty-first century by enabling users to see, edit, set deadlines, and cross items off all from within the web browser.

The extension has task options including setting alerts, filing tasks into folders, and flagging important items. Communication managers can create folders with designated tasks for each staff member and set alerts for priorities. Using Any.do, colleagues have a coordinated view of communication deadlines and can efficiently convey completion of tasks. This keeps everyone on the same page and helps hold team members accountable for their respective duties.

Black Menu

For me, desk clutter tends to induce more stress and having multiple tabs and/or windows open is just digital mess. Assuming that you take this chapter's advice and use Google for your communication functions, the Black Menu extension makes navigating among the various Google apps smooth and clean. The extension button uses a drop-down menu that is accessible from any browser window. Hover over the apps in the list to get a preview of stored content and to directly manage it. For instance, I can hover over the calendar to get a preview of upcoming events and search for those further into the future without fiddling with another window tab. To open any app, click on it in the menu and it will open into a new web page.

Client for Google Analytics

Tracking traffic to your site is a key component to communication practice. Every library should be aware of where their website's users are coming from, which web pages they are viewing, and how long they are hanging out on those pages. Google Analytics is a free way to monitor this. Setting up an account and embedding its code in your website allows you to view a multitude of statistics and create monthly reports.

For those library staff members who compulsively check these things, Client for Google Analytics enables a quick view of a page's statistics. Instead of logging in to a Google Analytics account, you simply click on the extension's button and instantly view data. This ability is crucial for efficiency when monitoring the launch of a new web page.

Hootlet

Statistics generally show a higher return on investment for social media pursuits than traditional outreach such as submitting press releases and purchasing TV ads. Having multiple well-managed social media accounts is crucial. A tool called Hootsuite allows one or multiple account users to manage up to three social media accounts for free. Using their online dashboard at hootsuite.com, staff can schedule posts to Facebook, Twitter, Linke-dIn, and more from one window. Hootsuite also enables users to respond to messages and track posts to analyze their outreach, among other great features.

Google Chrome makes Hootsuite all the more worthwhile with their Hootlet extension. This extension allows you to create posts without going to the Hootsuite dashboard. You can copy links from current web tabs directly to one or all of your social media accounts. The extension also offers the same flexibility as the dashboard, with options to schedule posts, add attachments, set a location, and so on. Using Hootlet, seamlessly integrating community news into your social media accounts is simpler than ever.

With all the potential content out there to share, I find it hard to remember something notable that I read a day ago or even last week. The Hootlet extension saves my sanity. When I see something that is of interest to our library community, I schedule it and forget about it. This makes it easy to keep a social media account fresh without spending each day contemplating what to post.

TinyURL

Most library staff are familiar withTinyURL.com. It is a great tool for sharing links while providing information for a reference query or posting to social media. By installing the extension, there is no need to go to the TinyURL website. When you're on a page that you want to grab a shortened link for, select the TinyURL button and voilà! Just copy and paste to easily share content within an e-mail or social media platform.

Save to Google Drive

As discussed earlier, using one shared profile among library staff involved with public relations enables them to more easily share documents via Google Drive. The Save to Google Drive extension enhances this participatory method by making it easy to save documents, images, videos, and links. Simply click on the extension button or right-click on the content you wish to save.

These gathered items can be renamed and/or stored in folders assigned to specific staff members. Similar to using Any.do, assigning projects to staff using folders within a shared Google Drive helps keep track of who is doing what. At the same time, it allows everyone to view each other's contributions for possible constructive input.

vidIQ Vision for YouTube

The high usage of YouTube's search engine by web users to find resources is clear. Libraries can harness the power of this by developing their own YouTube channel and sharing content.

Similar to the Client for Google Analytics, vidIQ enables you to analyze video viewings on YouTube. Once the extension is installed, the statistics automatically pop up in the top-right corner of a YouTube video's web page. It provides data such as the number of views, Facebook likes and shares, and YouTube comments, among others. It also gives tips on improving the video's outreach, such as having a longer description for better search engine optimization (SEO).

Our local-history supervisor recently uploaded some videos from our local-history collection to YouTube. This tool will help me evaluate what we can improve to better promote this material. Once the descriptions and keywords are better developed for each video, I will begin sharing links to the videos and publicizing them widely. The vidIQ extension will help me monitor statistics on views along the way and make revisions where they are needed.

CONCLUSION

Having a virtual presence was an optional advantage a decade ago. With most businesses moving their PR pushes online, it is now imperative to have an online existence and strategy. Virtual PR is theoretically a free way to promote your library's brand, but time is money, and monitoring your library's reputation can be tedious. Google products are practical tools to help with this workload. Even staff with no background in the communication field can find themselves becoming savvy contributors to their library's communication efforts.

The profitability of using these methods is rarely immediate, however. Similar to building outreach ventures for program and service partnerships, relationships with local media and the community take time to develop. At least one staff member should pursue the monitoring of your library's brand by scouring online content and trends. It may seem tedious, but your library will get a higher rate of return in the long run.

FURTHER READING

Fitch, B. 2012. *Media Relations Handbook: For Government, Associations, Nonprofits, and Elected Officials.* 2nd ed. Alexandria, VA: TheCapitol.Net, Inc.
Goldman, J. 2013. *Going Social: Excite Customers, Generate Buzz, and Energize Your Brand with the Power of Social Media.* New York: American Management Association.
Qualman, E. 2013. *Socialnomics: How Social Media Transforms the Way We Live and Do Business.* 2nd ed. Hoboken, NJ: John Wiley & Sons.
Seitel, F. P., and J. Doorley. 2012. *Rethinking Reputation: How PR Trumps Marketing and Advertising in the New Media World.* New York: Palgrave Macmillan.

Chapter Twenty-Seven

Google Drive for Library Productivity

Teri Oaks Gallaway and Jennifer Starkey

INTRODUCTION AND OVERVIEW

Collaboration and shared decision making are hallmarks of the library profession, and Google Drive is a productivity tool that facilitates this approach to work. The variety of functions available through Google Drive support library staff as they seek to streamline and simplify workflows, routines, and approaches to information sharing. This chapter presents a menu of practical uses of Google Drive for library staff communication and assessment, including strategies for real-time sharing of metrics and data, spreadsheets for project management, collaborative approaches to workgroups and committee work, staff portfolios and dossiers, and more.

Libraries generate a vast amount of data through a variety of systems, websites, door counts, usage statistics, and circulation statistics, just to name a few. That data is often stored in unconnected systems and may be challenging to access and correlate. The people who have the expertise to retrieve and format that data are not always the same people who make the data-driven decisions. In this chapter are several examples of how Google Drive can bridge these gaps for successful shared assessment projects. Google Drive is also an excellent tool for enhancing productivity in groups, supporting collaborative processes such as writing, brainstorming, decision making, conducting surveys, and managing large data projects. It is not surprising that so many organizations have embraced Google Drive as a primary workspace.

CORRELATING CIRCULATION STATISTICS, GATE COUNTS, AND WEBSITE TRAFFIC

In times of limited resources, a library director may ask: When should the library be open? Are there times of the day when the library could reduce its staff, open later, or close earlier? Using data to answer these questions can be challenging because the information that is needed may be difficult to access without systems administration expertise. Granting access to the gamut of data available in library systems may also introduce privacy and security issues. Using Google Drive, the systems experts can extract that data from the catalog in tab-delimited or spreadsheet format and share it with multiple stakeholders. The data is then easy to share and safely manipulate without compromising the security or integrity of the native system.

What if the format of the data in your native systems aren't easy to import to spreadsheets? This could become a great student-worker data conversion project. Create a Google Form for your students and see how fast your gate counts, website statistics, and circulation by hour statistics will come together in one spreadsheet. Here you'll be able to easily sort your rows to find out how busy the library is on a Tuesday at 6:00 p.m. There is no more having to wait for systems experts to create a detailed report and analyze it. If you incorporate data from multiple years, you'll be able to answer questions such as whether it is necessary to stay open late the day before Good Friday.

Tips

- Instead of entering data into a spreadsheet, create a Google form and link it to a Google sheet.
- Create a system for reviewing student-entered data. Do the transactions per hour add up to the right daily total? Create a formula to add them up and check.
- If data is entered day by day, print a calendar (or use a shared Google Calendar) to check off days as they are entered and allow for simultaneous data entry.
- Include a column for the day of the week, for each hour of the day, and to specify if a day is a holiday.

REVIEWING DATABASE AND SERIALS SUBSCRIPTIONS

Many constituents are involved in decisions to cancel or add databases or journal subscriptions, and not every institution or organization has access to an electronic resource management (ERM) system, which makes this process easy. Additionally, it may not always be best to invest in training multiple selectors, faculty, or content experts on how to access and interpret the data that is available through those systems. A Google sheet can enable multiple stakeholders to view, edit, and sort data that has been harvested from usage reports, subscription agents, link resolvers, or an A-to-Z list. It can be edited on the fly to include questions or notes about titles or subscriptions, decisions to keep or cancel, and notes about things like interlibrary loan requests, unique coverage, or embargoes. Up-to-date information can also be provided to nonlibrary staff for grant applications, program reviews, or accreditation purposes.

Tips

- Always make sure to include a unique identifier, such as an ISSN, to avoid improper title-sorting issues.
- Include a column for a subject area or discipline to allow easy sorting.
- When rows contain too many columns to be visible in one screen, use "freeze rows" to keep the sheet readable.
- Bring your print journal circulation or in-house usage reports together with your electronic journal usage statistics to create a multiyear picture of the use of your entire journal collection.
- Experiment with using the built-in pivot table function to create your own overlap analysis from vendors' title lists.
- Let selectors use a Google sheet to rank and prioritize purchases.
- Add columns to track annual prices, and use a formula to calculate your annual inflation.

DEACCESSIONING PROJECTS

Deaccessioning projects present difficult workflow challenges. Early in the process, your workgroup should decide whether selectors will look up circulation information from the catalog with a laptop in the stacks or view prepared reports. If prepared reports are preferred, use this as an opportunity to export your circulation reports to a Google sheet. This way, all participants involved can have an online conversation about which books should stay or go. This will solve problems caused by hand-marked shelf reports and will obviate the need to flip books on their spines, which always raises concerns that they will be reoriented before they are weeded.

Once the items to be deaccessioned are identified, many steps must be completed. Were the volumes removed from the shelf? Are they being offered up for donation? If so, to whom? Have records been removed from the catalog and from OCLC? To complicate matters, it is unlikely that one person is involved in every stage of managing the deaccession, so good communication is essential. After carefully considering all the steps to complete a weeding project, consider translating those discrete steps into a shared spreadsheet to which all parties can contribute.

Tips

- Ask selectors what statistics or data would be helpful to them, and consider adding a column such as number of OCLC holdings.
- Include an item barcode or other unique number in addition to a call number.
- Use row or column color coding or text style to allow quick scanning for status.
- Include a column to add a reason for the deaccession and supply a controlled list of reasons. This could be helpful if decisions are questioned or if multiple selectors have jurisdiction over the same areas of the collection.

COLLABORATIVE BRAINSTORMING

Libraries value the input of staff from all areas in the development of strategic-planning documents; team or committee plans; and mission, vision, and values statements. Additionally, collaborative group processes can keep the work interesting and productive. Making use of Google Drive for both synchronous and asynchronous divergent-thinking activities allows many points of view to be included in an easily manageable mode. Furthermore, the asynchronous mode that Google Drive offers can facilitate contributions from less vociferous staff and dampen the effects of the more aggressive participants. Another approach to collaborative brainstorming for groups is to create a set of questions or prompts in a Google doc, share the document with members of the group with some time for individual thought, then project the document onto a screen while one person records. This could also take the form of having multiple small groups contributing to the Google doc using laptops. This method can replace traditional whiteboard or flip chart recording, which requires transcription and organization afterward.

Tips

- Use think, pair, and share activities to facilitate the brainstorming process and record the ideas generated in a shared Google doc.
- Use the comment feature to add suggestions for revisions that the whole group can see before the edits are accepted or rejected.

- Use a speakerphone to bring in remote participants. They can open the Google doc on their own computers and create a screen-sharing environment in which to brainstorm.

COLLABORATIVE WRITING

Producing a document with a group can be a difficult process to navigate. Just like group brainstorming, a Google doc can facilitate a more productive experience where multiple contributors can have their voices heard. In co-writing situations, individuals may be assigned to work on different sections of a document, but by working on a shared document authors have the opportunity to see one another's work develop. This assists the authors with maintaining a consistent style and in avoiding work redundancy. Authors can also easily work together at a distance via conference call or the chat function built into Google Drive applications, all while viewing the shared document. It is far more efficient to edit a single shared draft in real time than to create multiple versions of a document; it can be challenging to keep track of which draft is the most recent version and whose turn it is to edit the single authoritative copy.

Tips

- Determine the final editorial process to assure consistency and to review comments and incorporate changes.
- Use highlighting or comments to mark areas in the document where input or revision is needed.
- Use the revision history function to revisit previous versions if needed.

LIBRARY PROGRAMS/INSTRUCTION TRACKING

The teaching role of academic libraries and outreach and public program initiatives at public libraries are priority efforts that deserve quality documentation and analysis. Google Drive can play a role in the way a staff communicates about instruction/programming and in understanding trends in the needs of users, such as changes in the curriculum or popularity of public library programs. Additionally, keeping track of instruction or programming in this manner facilitates agile data-driven approaches to decision making.

A well-organized spreadsheet can provide rich data at a glance. Which academic departments are using instructional services heavily, and which academic programs present opportunities for library outreach? How is the instruction load distributed among teaching librarians? How many students are served by the instructional program? Where does instruction take place?

What is the format of the instruction session (presentation, activity, lab session, flipped class, etc.)? Similarly, public libraries can answer questions such as which types of programs have the highest participation? Are there gaps in the programming, populations that are served, or topics that have yet to be addressed?

Tips

- Consult with stakeholders to determine what your organization wants to track, including special questions such as course instructor attendance or event location.
- Develop data standardization, such as course name/number format and representation of classes with multiple library sessions.
- Consider using a Google form to populate a Google sheet.

REFERENCE SERVICE TRACKING

There are many commercial options for reference transaction data keeping (LibAnswers, libstats, Gimlet, and more), but a Google sheet is a free and easy way to collect reference question statistics. Questions can be logged directly into a spreadsheet or using a Google form connected with a Google sheet. Reference transaction data can be used to identify peak times of the year and busy or slow hours of the day to inform desk staffing decisions. It can identify trends in the amount of time spent answering a question, the nature of the questions, simple or complex research questions, and directional or technology questions. It can also be used to help train new staff by providing real examples of questions asked.

Using a Google sheet to keep data for research consultation programs in academic libraries can also be illuminating in different ways. Tracking the numbers of the courses for which students request research consultations can be used to inform instructional programs. Are there courses that generate a larger number of appointments? Did those courses have instruction? Are particular faculty members recommending or requiring research consultations? Are there assignments that are problematic for students or that students are completing at the last minute? And how can the library better engage the faculty?

Tips

- Google Sheets has less data analysis functionality than Excel, and you might want to download Google data as an .xlsx file to work with in Excel.

- If multiple stakeholders are using and sorting data in the same document, it can become problematic; doing sorting and analysis in Excel can safeguard files against mistakes.
- Decide in advance what data you would like to collect and what you plan to use it for, such as whether you wish to capture the content of the questions for a knowledge base.
- Always consider patron privacy when logging individual interactions; avoid using the names or other identifying information of patrons.

SURVEYS

Google Drive can be used to create surveys of any type, from large-scale user surveys to small or informal surveys of library staff. It is a free alternative to subscription-based tools such as Zoomerang or SurveyMonkey, and it is very simple and straightforward to use. There are a few drawbacks to using Drive instead of for-pay survey tools. The data-analysis tools available in the Google Sheets application are not as sophisticated as those built into for-pay survey products, but data can easily be converted to an Excel file for more advanced analysis. Keep in mind that Google Forms cannot support surveys with internal logic (e.g., if a respondent answers "no" to question #3, skip to question #5). Google Forms does allow creators to divide surveys into multiple pages and provides several options for question types, including multiple choice, text boxes, and checkboxes. However, Drive does not enable "ordered ranking" questions or a matrix-style question allowing respondents to answer several questions using the same Likert Scale, for example. With a few exceptions, Google Drive is an excellent tool for most types of surveys that libraries deploy.

Tips

- First consider whether Google functions are sufficient for the type of survey you are conducting.
- Decide who should have access to the data, and use the privacy settings to control access.
- Decide where you want the data to be stored by using the function "choose response destination" at the top of the Google Forms editing menu. If you do not select a spreadsheet, Drive will create one in Google Forms.

STAFF PORTFOLIOS AND DOSSIERS

Many professional librarians undergo periodic reviews, performance appraisals, or evaluation for tenure. Since library work is increasingly represented in digital formats, having an online portfolio can solve problems of representing digital professional output in a paper format. An individual compiling a dossier can use a Google Drive folder to store digital copies of files such as a resume or curriculum vitae, slides of presentations, publications written, and other digital documents and professional contributions.

Tips

- Be sure to adjust the sharing settings so that dossier recipients work with read-only files.
- Use subfolders to group-related materials if there are more than a few files.
- Upload digital copies of printed conference programs or certificates.

WORKGROUPS, COMMITTEES, AND PROJECT MANAGEMENT

Google Drive is changing the way work gets done in many organizations; it has simplified storing and sharing of workgroup documents, minutes, agendas, and files. A member of a committee can easily create a folder to share with the Google accounts of other members (the folder appears in the collaborators' "shared with me" section on the left-side menu in Drive). Alternatively, members of a group can have a single shared Google account that all users can log in to simultaneously. Shared folders can have subfolders, and sharing settings can be adjusted for each individual document or subfolder.

Another innovative approach to committee and workgroup management is to use Drive to create and store agendas and minutes. It can be a very efficient and democratic approach to agenda-setting because all stakeholders can add agenda items directly to the document. The minute-taker can make notes directly on the agenda document, which saves time and provides the added benefit that participants can see the minutes expanding in real time, which helps assure that minutes accurately reflect conversations and decisions. Using Drive to archive past minutes also makes it unnecessary for all members to keep local files of minutes.

Tips

- Drive can also store many other document types such as Microsoft Office files, PDF files, and some image types.
- Drive is free for up to 100 GB of storage; prices in 2014 are $1.99/month for 100+ GB, and $9.99/month for a 1+ TB of storage.
- Keep in mind that there are file-size limits that vary by file type.

- Files can be converted back and forth between Google and Microsoft applications.

GOOGLE DRIVE ADD-ONS

In addition to the core functions provided by Google Drive, a set of add-ons and extensions is available and growing, and most are free. These add-ons can be easily downloaded and used in your existing Google Docs and Sheets. While some of these are built-in functions of existing versions of Microsoft Office products, several are unique to Google Docs. A sampling of currently available add-ons are described below:

- *EasyBib Bibliography Creator*. Easily format citations for your documents. A powerful search interface allows authors to search for citations using keywords, author names, ISSN number, or a DOI and then instantly format and embed those citations in a bibliography. Even though a limited number of citation styles are available, this tool is so intuitive that students and librarians alike are sure to embrace it.
- *TextHelp Study Skills*. A sophisticated highlighting tool that allows users to extract key points for study. Using color coding, it is easy to locate passages of a document to include in a summary document. This tool would easily facilitate the creation of an executive summary of a team or committee report.
- *Table of Contents*. By applying built-in paragraph styles (e.g., Header 1, Header 2) a dynamically formatted table of contents can easily be incorporated into any document. Make a change to the text or organization and it updates automatically when the document is reloaded or refreshed. Once the section headings are created, it is easy to navigate through the document to a specific section by clicking on the section link in the table of contents sidebar.
- *Track Changes*. This feature has been long available in Microsoft Word and was an obvious missing element for effective group collaboration in Google Drive. A summary of all changes, listed by username, is shown in the sidebar with easy tools to accept or reject the changes.
- *LucidChart Diagrams*. An extension that allows authors to easily create and embed standard diagrams, flowcharts, organizational charts, and floor plans into a Google doc. Most of the templates are free with the exception of templates for wireframes, Visio imports, and a few others.
- *ProWritingAid*. Although this is not a completely free add-on, it is especially innovative. A consistency report analyzes elements including hyphenation and capitalization. Other reports are available that check for properly defined acronyms, use of clichés, redundant language, plagiar-

ism, and grammar. The first four reports are free, as is a fourteen-day trial, after which an annual license can be purchased for a small fee.

- *Gliffy*. This tool is very similar to LucidChart and similarly offers both free and paid templates. Gliffy has been available as a free, stand-alone diagramming tool for quite some time. An existing account can be linked to a Google Drive account to allow for easy importing.
- *UberConference*. Hold the phone! This is a free built-in conference call extension that allows document collaborators to participate in a conference call using either their built-in computer microphone and speakers or a phone. A sidebar in the Google doc shows which participants are currently participating in the call. Conversations can be recorded, and up to ten participants can join the call with the free subscription. The add-on is very similar to the functionality of Adobe Connect.
- *SuperMetrics*. A tool to bring in existing data sources such as Google Analytics, AdWords, Facebook, and Twitter to your Google docs. Data can be formatted with standard visual tools including tables, charts, and graphs.

FURTHER CONSIDERATIONS

This chapter presents an array of productivity ideas for collaborative workflow in libraries. For a more detailed discussion of Google Drive functionality and a critical analysis of its capabilities, see our review in the *Charleston Advisor* (Gallaway and Starkey 2013). Additionally, the review explains how Google Drive has the ability to synchronize and back up files to a folder on a computer hard drive. All users of Google Drive should take care to use the local folder or some other means of backing up files in Drive. Many organizations have adopted Google Drive with great success; in fact, the Google Apps for Education initiative has made it easy for educational organizations to provide information technology functions on a locally branded Google platform. While some organizations have embraced Google for communications, organizations where sensitive information is at stake have been cautious, claiming security concerns. With the recent introduction of add-ons, new functions, both free and paid, will likely be added at a rapid pace, giving users even more incentive to consider the collaborative and assessment possibilities in the Google Drive environment.

REFERENCE

Gallaway, Teri Oaks, and Jennifer Starkey. 2013. "Google Drive." *Charleston Advisor* 14 (3): 16–19. doi:10.5260/chara.14.3.16.

Chapter Twenty-Eight

Google Translating and Image Searching for Foreign-Language Cataloging

Laura Bohuski

The rise of the Internet has changed the world in ways that we have not even begun to comprehend. The impact of the World Wide Web on the function of libraries has been so significant as to change the very foundation of how libraries and librarians work. Google has taken the changes even further, pulling patrons away from the library. Libraries have had to adapt to survive, and cataloging has had to adapt as well.

Before the Internet search engines simplified information retrieval, cataloging was mainly restricted to the cataloging of hard-copy materials, the knowledge in catalogers' heads, and what catalogers could learn from cataloging manuals. Cataloging foreign-language materials was difficult even with translation aids such as a Japanese to English dictionary. Even though Google has changed how libraries function, the creation of Google Image Search and Google Translate has made the cataloging of foreign-language materials much easier.

For many English-speaking catalogers, cataloging foreign-language materials is hard enough when the foreign language uses characters from the Roman alphabet. Cataloging becomes more complicated when the language is written in a character set completely unrelated to the Roman alphabet, such as hiragana in Japanese, simplified Chinese characters, or Cyrillic characters from Slavic languages. Finding ways to translate foreign titles into characters that a cataloger can use to search English-language databases takes some ingenuity and forethought.

This chapter describes the ways I used Google Image Search and Google Translate to successfully translate multiple Japanese-language titles into both the Roman alphabet versions of their native language and into English. For the purposes of this chapter, I will be using two books written in Japanese hiragana, katakana, and kanji from a collection that was given to Western Kentucky University's library. While the chapter focuses on translating Japanese to English, some or all of these steps can be used with any foreign-language translation.

STEP ONE: IMAGE SEARCHING

Some cataloging systems cannot read non-Roman characters and cannot display them in the library's database search engines. Such is the case where there is no input method available with my integrated library system for non-Roman characters. Finding a way to re-create the title of any foreign-language material I am cataloging in such a way as to be able to create a string of non-Roman characters took some thought. I know some Japanese, so I am aware that charts of basic Japanese characters exist on the web. Finding the right ones is just a matter of time and patience.

But I also know that finding the more complex characters, such as Japanese kanji and Chinese characters, is much more difficult. Searching just for individual characters will not necessarily produce the book title in its entirety. In cases where significant portions of the title are written in kanji, you can spend days searching for the correct character, especially since there are over fifty thousand kanji characters in use with approximately three thousand used commonly. While a list of common kanji and their translations can be found through a basic Google search, taking the time to search through each one is daunting, especially when there are faster ways. I will further explain the two methods I found to search for the more complex characters using Google Image Search before moving to the next step, which involves Google Translate.

Step 1.A: Character Searching

The best way to acquire a correct translation is to place the exact string of characters from the title page into Google Translate. The most efficient way I have discovered to find these characters is by character searching. Character searching is the simpler of the two types of image searching I engaged in to create English titles from foreign-language materials. Character searching is exactly what it sounds like; you search for a chart of written symbols and find the characters that match the characters on the book. Luckily for catalogers, books coming into the library often have some identifying information linked to them. The language the book is written in is one of these pieces of information, and this starting point can be all that is needed.

When I first started using this method to translate Japanese titles into the Roman alphabet, I was striving to transliterate them into romaji, the Roman form of Japanese characters. Since I had studied some Japanese, I thought that obtaining the romaji title first would be the best way to go about the transliterations. I could use my knowledge of basic Japanese words and Internet searches to translate the titles easily. But there were some issues with this process.

First, when the words were written in romaji instead of hiragana or kata-kana, it was difficult to find a trustworthy translation. Searching for each individual string of symbols was also very slow. Another factor to consider is that certain combinations of katakana and hiragana form different words and colloquial phrases. These combinations are difficult to search for, especially when the combinations can change with the addition or subtraction of a single character. Add in the kanji and there is a whole new collection of phrases that are inordinately difficult to search for.

Second, when I ran a string of romaji through Google Translate, I realized that it cannot translate romaji into English; it only translates Japanese charac-ters. So, while having the title in romaji was nice, it wasn't necessary.

The first book to be cataloged has this string of characters on the title page: "日本の歴史 17 町人の実力" (Naramoto 2005). I could tell from the composition of the book that this was most likely a volume in a series. From the image on the cover, it looked like there might be content about ancient Japanese culture or geisha. I could also assume that the text after the number 17 was a subtitle.

Of the characters in the title, two are exactly the same. The Japanese hiragana の, which spells "no" in romaji and usually means "of," and the katakana カ, which translates into the roman letters "ka," which is one of the basic Japanese alphabet symbols. The rest of the characters are more difficult to find since they are kanji. A quick search for "common Japanese kanji" produces multiple lists to choose from, and many that are easy to peruse.

The website I used to find kanji characters is http://tangorin.com/common_kanji (Tangorin 2014). This website has lists of the basic common kanji needed for language fluency from grade school through all five levels of the Japanese Language Proficiency Test. Most important, the individual characters on this list can be copied and pasted into a word document.

When a cataloger finds a set of lists that works well, the meticulous part of this process begins. The best way I have found to locate these characters is to carefully look through each list. When beginning to catalog foreign-lan-guage materials in an unfamiliar writing system, taking your time in the beginning will save time later. Start compiling a list of common kanji with their meanings to shorten the process of looking for repeat characters.

Once you have begun assembling multiple characters, you then copy and paste them into a text document in the same order as those displayed in the title. From there it is a simple task of copying the string of characters into Google Translate. Moving to that step depends on how many of the characters from the title have been found.

If, as with this first title, all of the characters from the initial searches have been found, then you can skip the upcoming book cover image search and move to using Google Translate. If you have a title that is more complicated and finding the complex characters is taking too long or becoming frustrating, your next step is to search for book covers.

Step 1.B: Book Cover Image Searching

Book cover image searching is similar to character searching, but is only useful once you have first attempted character searching. Google's advanced search features allow you to use symbols, such as asterisks, in place of letters or phrases during searching. For example, an asterisk can be used in the place of a missing letter, word, or phrase that you don't know or remember. Google interprets the asterisk as a wildcard or a blank space and returns the most likely results for that missing wildcard within the string of characters provided.

It's best to start with character searching since you will need at least a few of the characters to provide an initial search string. Searching for book covers is useful when there is a character or set of characters you cannot find or if you're looking to save time. The book I used for this demonstration has this string of characters: "図説民俗探訪事典" (Akio 1983). There are no basic hiragana or katakana characters in this title, and searching for each character through multiple lists can easily take a couple of hours. With book cover image searching, you can greatly reduce that time by first searching for a few of the title characters.

The number of characters needed depends on the uniqueness of the title and if the book image can be found on the Internet. Unfortunately, this technique does not usually work with plain book covers that have solid fabric covers or only text, because people are not interested in posting images of plain books. Though you are unlikely to get too many hits for a Japanese title, with a plain book cover you are very likely to get none at all.

When searching for a book cover, it is useful to use different ways to search for the characters needed to create a successful search string. The first method is to search each character list in order and locate any of the characters that are on those lists. You might have to search each list multiple times to find all the characters. With the example title shown above, that would include finding all but the fourth character, which sort of defeats the purpose of book cover searching. If you do use this method, the following search string, with the fourth character missing, 図説民*探訪事典, will return a link to the Amazon.com Japanese page for this specific book (Amazon.co.jp 2014).

Searching for characters as they appear in lists is a good method, but this means you have to find seven characters to effectively search for the book's image. Instead, if you search for the individual characters in the order they appear on the title page, you only have to find four characters before Google returns the same image. This search string, 図説民俗*, will also return the same image and web page with half the numbers of characters in half the time.

STEP TWO: TRANSLATE

There are different methods of using Google Translate. The first is using the translate website at http://translate.google.com (Google Translate 2014b). The second can be used when using Google Chrome as your web browser. Google Chrome has a translate feature built into the browser and can translate entire web pages. Other browsers, such as Internet Explorer and Firefox, do not have this feature. Both of these forms of Google Translate can be effectively used to translate Japanese characters into English titles.

The official Google Translate website can be used from any browser and is very user friendly. A text box that detects the language of the words placed inside it takes up one half of the screen, and the other half is an output box. Each box has a few tabs and drop-down menus above it where you can tab between common languages that you've translated to or from or pick a new language if needed. The detect feature works very well on the text box, but this is the feature where I learned about Google Translate not translating romaji. It detects romaji as Vietnamese or another East Asian language, and when I corrected the input language to Japanese, it returned the romaji exactly as I had entered it. Google Translate does not recognize the romaji as Japanese in any way, but it easily recognizes the Japanese characters.

Even when Google recognizes the language and translates, sometimes the translations that come out are incorrect or out of order. When the characters from the first title are placed into Google Translate, the resulting title is "Real mosquito 17 townspeople Japanese history." It really doesn't make much sense. The last four words have some meaning, though they are probably out of order, but "real mosquito" has no place in that title.

Google Translate is run by multiple algorithms in which a computer polls millions of documents that have already been translated by humans and then looks for patterns. These patterns are translated into the words in the output text box (Google Translate 2014a). When there are not as many documents already translated by human translators, there are not as many patterns for Google to reference, so the translation results do not always match what catalogers are predicting the titles to be. In these situations, a cataloger's discretion comes into play.

Usually when a translation like "real mosquito" is so obviously wrong, Google Translate offers multiple options for the translation based on the patterns the software has discovered. An example of this is the translation from the second book's cover image. The original translation of the title was "Illustrated encyclopedia explore folk." While this title works grammatically, it is much easier to see what a correct English title might be than with the first example. Also, Google Translate allows the cataloger to choose a different translation if needed.

Choosing different translations is simple. When you click on a word that you believe is incorrect, a drop-down list of other translations appears. The symbols that these words represent are highlighted to provide an easy correlation. Look at the list and pick a word that more correctly complements the other words in the title. The list also includes images of the characters that the algorithms are translating, allowing those with almost no familiarity with a foreign language to choose a better translation. In this case, Google Translate and English grammar help us to change the unwieldy "Illustrated encyclopedia explore folk" into "Illustrated encyclopedia exploring folk customs." While this translation is not completely correct, the title makes sense.

Having the ability to pick a different translation is only useful when a better translation is available. The book title with the beginning translation "real mosquito" does not have a "better" translation, which is unusual for Google Translate. Luckily for a cataloger, Google Translate provides a romaji translation of the characters in the text box. The romaji is in light gray text underneath the text input box, and a simple Google search will usually return a meaning for the romaji word. You can also switch characters around or delete characters to see how the words change in response and determine a title that works for your material.

Using the first book as an example, it is possible to isolate which kanji or alphabet characters are producing the strange translations. "Japanese history" is the translation for the first five characters in the title. This makes sense in the context of the book, and so does the number 17, because I know it is the seventeenth book in a series. With those characters deleted from the text box, we still get a translation of "real mosquitoes townspeople." "Townspeople" is linked to the first two characters after the number 17, "町人." Removing one or the other or both just changes the word those characters represent. So the issue is with the last three characters, "の実力."

Using Google Translate's ability to highlight shows that the translation "real" comes from the kanji in the center, or "jitsu," and that the translation of "mosquitoes" is from the combination of the first character, "no," and the last character, "ka." We know that "mosquitoes" does not fit with the rest of the title, so it is logical to search for another word meaning.

As stated earlier, the Japanese character for "no" can be translated as "of" or as a possessive such as "of the." The katakana and kanji form of "ka" mean "ability" or "power." This can be found by searching the symbols of "ka" and "no" separately on the Google Translate page. They are separated in the title by another character, so they are probably not meant to be translated together in this situation.

With these other meanings found and with some common sense, we can create a subtitle that works better with the series title "Japanese history." If we transcribe the character translations exactly, the title that forms is "Townspeople of the real ability." If we switch it around so the subtitle works with the rules of English grammar, we achieve "Japanese history 17: Real ability of the townspeople." This title makes much more sense than "real mosquitoes" and, when searched for in databases like OCLC, will return more titles.

STEP THREE: CATALOGING

This is now the easy/fun/exciting part. By now you should have at least two versions of the title, one in English and one in Japanese characters. If you thought to copy and paste the romaji from below the text box of the Google Translate page, you might even have three separate versions of the title. It has been my experience that a search in a database such as OCLC will return a record for at least one version of this title that closely matches the title in your hand. The records in OCLC use Japanese symbols, romaji, and English. I check the symbols from the title on the material against the records in OCLC. I find the one that best matches my book and then import that record to my integrated library system. I then modify anything I need to in order to make the record work within my system, such as deleting Japanese characters, and then holdings and item records.

Once I learned how to re-create an original book cover title, I haven't needed to create original bibliographic records often. When I find that I need to create an original bibliographic record for foreign-language material, I search the complete title in Google Chrome and hope to find a web page with the book's information. I use Chrome's ability to translate the web page to retrieve information such as the author and publisher, if possible, and then use the physical descriptors of my book to fill in the rest. Cataloging foreign-language materials can prove frustrating, but using the steps above can resolve most of that frustration and provide access to titles that increase the value of the library's collection.

Cataloging can be difficult at the best of times; cataloging foreign-language materials in a language that you cannot read or understand just increases that difficulty. By using Google Image Search and then Google

Translate to compile the work's original title, a cataloger can greatly decrease the time and frustration spent on each record. While the introduction of Google into the world of libraries has intrinsically changed the way libraries function, librarians can, in turn, use Google to enhance their own work and make Google work for them.

REFERENCES

Akio, Oshima. 1983. 図説民俗探訪事典 [Illustrated encyclopedia exploring folk customs]. Tokyo: Yamakawa Publishing.
Amazon.co.jp. 2014. "図説民俗探訪事典 [新書]." Accessed January 20, 2014. http://www.amazon.co.jp/図説民俗探訪事典-大島-暁雄/dp/4634600900/ref=pd_sim_b_3.
Google. 2014. "Chrome." Accessed January 20, 2014. https://www.google.com/intl/en/chrome/.
Google Translate. 2014a. "About Google Translate." Accessed January 20, 2014. http://translate.google.com/about/intl/en_ALL/.
———. 2014b. "Google Translate: Japanese to English." Accessed January 27, 2014. http://translate.google.com/#ja/en/.
Naramoto, Tatsuya. 2005. 日本の歴史 17 町人の実力 [Japanese history, vol. 17, Ability of the townspeople]. Tokyo: Chuo Koron Shinsha.
Tangorin. 2014. "Common Kanji." Accessed January 17, 2014. http://tangorin.com/common_kanji.
Wikipedia. 2014. "Hiragana." Last modified January 14, 2014. http://en.wikipedia.org/wiki/Hiragana.

Chapter Twenty-Nine

Understanding Our Users

Feedback and Quizzes Using Google Forms

Josh Sahib and Mark Daniel Robison

INTRODUCTION: ASSESSMENT AND THE LIBRARY

The calls for more assessment in academic libraries seem to be growing louder all the time. From library administrators, provosts, accrediting bodies, and elsewhere come the directives to measure and record, to prove the value of academic libraries as higher education faces change on many fronts. Academic libraries have been proactive with assessment, especially in recent memory. Askew and Theodore-Shusta write that assessment has been a "growing subspecialty area in librarianship" in recent years, citing the growing number of conferences, institutes, and positions focusing on assessment in the last decade (2014, 1).

Assessment is driven by a need for accountability. Stakeholders want to know that the library's budget is being used effectively. Especially in light of the budget cuts that academic libraries have faced in recent years, library professionals need to demonstrate that their collections and services are achieving the intended outcome: vitally contributing to their campuses' academic missions.

All assessment methods are not created equal. Assessment data can be either quantitative or qualitative, but it also comes in a range of depth. Libraries excel at traditional surface-level usage data, such as gate counts and circulation statistics. While this type of "outputs" data is simplest to gather and useful for comparing institutions, it is often superficial in addressing the question of whether libraries are truly adding value to the teaching, research, and learning happening on campus. The two higher-level assessment categories include performative and authentic measures (Whitlock and Nanavati

2013, 35–36). Performative assessment evaluates a student's comprehension of material by giving the student a task to perform. The student's ability to complete the task indicates how effective the instruction is. Authentic assessment is an extension of performative, taking a long-range look at progress— for example, evaluating a student's growth throughout his or her college career.

Instruction librarians have become quite familiar with performative assessment, asking students to demonstrate their knowledge by completing worksheets or assignments as part of the library instruction session. Authentic assessment, on the other hand, is much more elusive to academic institutions. These types of assessment projects not only require much more planning but also raise issues of patron privacy and gaining buy-in from multiple levels of the institution.

As libraries have moved to capture data for assessment purposes, many corporations have stood to earn considerable sums of money. From paid websites that aid with survey design to software that tracks reference and instruction transactions, many of the solutions used in academic libraries are proprietary (Angell 2013, 596). Budget cuts in recent years have forced many libraries to reduce staffing and collections expenditures. At some institutions, the use of freely available tools could provide relief from these budget constraints.

Google Forms provides a free alternative for gathering many types of assessment data. Forms are one of the document types creatable using Google Drive (formerly Google Docs), Google's cloud-based productivity suite. Using Google, libraries can build sleek, web-based forms to gather qualitative and quantitative input from patrons and staff. The data are stored in a responses spreadsheet, where they can be analyzed as needed. Using Google Forms to collect this information is an alternative to the paper-based questionnaires and assignments that many librarians still use to measure learning in the classroom or to collect patron feedback on workshops or other events. Born-digital feedback data is more convenient to review and less likely to delay the assessment process (or to be lost entirely) than paper-based data. This chapter reviews the implementation of Google Forms for assessment at the University of Alabama Libraries and offers practical instructions on how these methods may be replicated at other institutions easily and effectively.

IMPLEMENTATION AT THE UNIVERSITY OF ALABAMA

The University of Alabama Libraries' Gorgas Information Services Department began experimenting with Google Forms as a method for collecting patron feedback in 2010. The department's use of Google Forms has grown organically in the years since, as its librarians have applied the tool to different purposes as new assessment needs have arisen. The department handles reference and most information-literacy instruction for Gorgas Library, the main library on campus. Using the formerly named Google Docs, members of the department first created a form to gather feedback about its virtual reference service, Ask-a-Librarian. The brief survey asked whether the patron felt satisfied with the service received and allowed space for open-ended comments. A link to this survey was included in the e-mails sent by librarians in providing the service.

This technique was soon applied to the department's instruction program. Librarians developed a basic survey using Google Forms to gauge the effectiveness of instruction, requesting that students complete the online form at the end of sessions. The feedback form asked the respondents to indicate one useful thing they had learned and something that still confused them, along with the name of the librarian who had led the session. The responses were stored in a spreadsheet within the departmental Google Docs account, allowing librarians to review the patrons' answers and identify strengths and weaknesses in their lesson plans.

The department greatly expanded upon this instruction feedback concept in the fall of 2012, when it developed a robust instructional approach for the English department's first-year writing program. Under the leadership of its first-year experience librarian, the department adopted a two-shot instruction model. Librarians began holding two instruction sessions with nearly every English composition section on campus with the objective of introducing the students to fundamental information-literacy concepts, including keyword selection, construction of search strings, and source evaluation. This two-shot model presented the department with unprecedented amounts of class time with incoming students and new opportunities to assess student learning. Using Google Forms, the department created two quizzes to measure how well students had learned the concepts at the end of the class. Because the sessions used a course LibGuide as the core learning object, the librarians chose to embed these Google Forms quizzes directly into the LibGuide. This integration spared students from getting lost when navigating to the quiz. The quizzes used a combination of multiple-choice and free-response questions to test students' understanding. The Google Forms quizzes also included required fields, such as instructor name and course section, so that the responses, though anonymous, could be associated with a particular session and librarian. Because the department shares a Google Drive account, the results were tabulated in a spreadsheet that was accessible to all of the department's librarians and relevant administrators. This feedback data has been crucial in improving the department's instructional approach and in demonstrating student learning to stakeholders.

As part of its curriculum for the English composition instruction sessions, the librarians adopted a "flipped classroom" model. Students were asked to listen to several podcasts prior to their library session so that they would be better equipped to retain the information covered. The department experimented with different methods for determining whether students actually had listened to the podcasts in advance, including a paper-based worksheet and a quiz constructed in the university's course management system, Blackboard. Ultimately, the librarians created podcast quizzes using Google Forms. In contrast to the quizzes embedded in the LibGuides, the podcast quizzes were designed as a series of pages, and students were made to answer each question correctly before proceeding to the next question. This design allowed students to receive immediate feedback on how they were doing. Because the only possible grade students could earn was a perfect score, the quiz has served more as an assurance that students have attained some of the primary podcast knowledge before attending the library session.

GETTING STARTED WITH GOOGLE FORMS

The remainder of this chapter provides practical instructions on implementing Google Forms in library settings, based on the examples referenced from the University of Alabama. While these instructions model the examples of an academic library, these assessment measures could be employed in any type of library environment. For the sake of clarity in these instructions, the term *readers* refers to library personnel and other professionals who are using Google Forms to create assessment instruments and *respondents* refers to the forms' intended end users.

Google Drive may be accessed online at http://www.google.com/drive . Readers who have existing accounts with any Google product (e.g., Gmail) may use those credentials to access Drive. To create a new Google account for Drive, readers should follow the Create an Account link.

Once logged in to Drive, readers may create a new form by clicking the Create button and choosing Form. Users should begin by giving their form a short, descriptive title. Although the title can be changed later in the design process, it is best not to leave the form untitled, for the sake of clarity. When the form is put into practice, users' responses automatically will populate an associated Google sheet with the same title, followed by "(Responses)." Thus, readers are highly encouraged not to change the name of the form once any responses have been received.

Another early step in the design process is to choose a theme for the form. For some library settings, a basic theme, such as "Magazine," might be most appropriate. For a school library or public library, some of the more colorful themes could provide the desired style.

ADDING AND EDITING QUESTIONS

Once the title and theme are selected, readers may begin adding questions. Google offers several basic question types, including:

- *text* (for short answers);
- *paragraph text* (for open-ended questions);
- *multiple choice* (for choosing only one answer);
- *checkboxes* (for choosing multiple answers); and
- *choose from a list* (for choosing only one answer from a menu).

These question types will be the foundation of most forms that libraries build for traditional and performative assessment measures. Google Forms also includes several other advanced question types, such as scale, grid, date, and time, which are not covered in this chapter.

When a new question is added, it will be in Edit Mode, appearing in an expanded view. In this mode, readers may edit a question's title and help text, the basic components that respondents will see on the live form. The default question type is multiple choice. Readers may change the question type easily by clicking the Question Type drop-down menu and choosing the desired option. For the question types that offer multiple possible answers (multiple choice, checkboxes, and choose from a list), readers will see additional fields where they may edit these options. When finished editing a question, readers may click the blue "Done" button, which will cause the question to collapse.

Throughout the design process, readers may easily edit, duplicate, delete, and reorder questions. The Edit, Duplicate, and Delete buttons can be found by hovering over a question within the form. The icons for Edit, Duplicate, and Delete are a pencil, two sheets of paper, and a trashcan, respectively. To reorder questions within the form, readers may hover over the desired question, which will cause the mouse pointer to change into perpendicular arrows. At this point, readers may drag and drop the question to its desired location. When a question is in Edit Mode, a cluster of dots appears in the top center of its frame. This cluster can be grabbed as an alternative method for reordering questions. Lastly, when a question is in Edit Mode, readers will see a Required Question checkbox at the bottom of its frame. Readers should check this box to make a particular question required; otherwise, it will be optional.

At any point in the design process, readers may see how their form will appear to respondents by clicking the View Live Form button near the top center of the form. When viewing the live form, readers may be tempted to experiment with the form by submitting trial responses. However, it is recommended that readers finalize a form as much as possible prior to entering any responses. As mentioned previously, form responses are automatically stored in a corresponding Google sheet, with each question and its responses assigned to a particular column of the spreadsheet. When the first response is submitted, the column order is permanently established. Perhaps counterintuitively, rearranging the questions within a form after a response has been submitted will no longer cause the corresponding columns to shift accordingly. It is, therefore, best to finalize the question order and other details prior to submitting responses.

At the bottom of the form, readers may edit the confirmation page, which displays the message that respondents will see after completing the form. This message is customizable and deserves some consideration. The page offers checkboxes to activate or disable additional features, such as allowing respondents to submit further responses or to view the responses given.

SHARING, DISTRIBUTION, AND MAINTENANCE

Readers likely will want to collaborate with colleagues on designing and implementing assessment forms. A form's privacy settings may be edited by navigating to the File menu and clicking Add Collaborators. (With some Google Drive document types, this option is alternatively labeled Share.) This will load the Sharing settings dialog box. Here readers may enter their desired collaborators' e-mail addresses and set their level of access. Within this dialog box, readers may also change the form's privacy settings. Under the heading Who Has Access is a hyperlink marked Change, which allows readers to adjust others' level of access to the form. It is generally recommended that the default settings not be adjusted and that only invited collaborators be given editing privileges.

Once it has been finalized, readers may distribute their form in a variety of ways. Clicking the blue Send Form button at the top of the page will provide readers with the permanent URL for the live form. The form may be distributed to respondents as desired by sharing that URL—for example, as a link in an e-mail or through social media. Alternatively, readers may embed the form within a web page with relatively basic HTML programming knowhow. Within the form, readers may select the File menu and click Embed. A dialog box will display the embed code, along with fields for adjusting the height and width of the inline frame ("iframe"). This code may be copied and pasted into the HTML code of a web page. Readers probably will need to adjust the height and width dimensions of the iframe code, so that the form fits neatly into the page. Even with little experience in web design, readers should find it straightforward to experiment with dimensions, using a trial-and-error system of adjusting height and width until the iframe is the right size.

When a form is not in use, readers might wish to deactivate it. Clicking the Accepting Responses button at the top of the form will deactivate the form, causing this setting to change to Not Accepting Responses. When a form is deactivated, readers may edit the error message that respondents see when trying to access the form. This error message may be customized to include additional instructions or contact information as needed.

If the responses are saved in a Google sheet that is accessible to multiple collaborators, there exists a risk of inadvertent data corruption. Because of Google Drive's versioning system, readers who notice errors in their data may restore previous versions of the sheet by clicking See Revision History under the File menu. From the Revision History menu, readers may select a previous, uncorrupted version to restore. Please note that users will lose any responses that were submitted after the chosen restore point. At the end of a semester, librarians may wish to export their response data for archival purposes. To perform this action, readers may simply open the File menu, click Download As, and choose the desired file format.

CREATING A FEEDBACK FORM

Readers providing instruction or other public services might wish to gather patron feedback through a Google form. This form could be embedded into a course guide, saved as a bookmark on the library's computers, or even loaded on an iPad that could be passed among participants. For the feedback to be of maximum usefulness in later assessment of services, it is best to keep responses anonymous but still include some relevant identifying information, such as the librarian's name or the course number. It is important to note that as each response is submitted, a timestamp is automatically captured with the response; thus, questions about the time of the session or service provided are unnecessary.

Next, readers may add the specific questions desired in the feedback. The questions asked will vary based on the use case and could range from simple multiple-choice questions to open-ended text questions using the Paragraph box type to allow respondents to develop their thoughts fully. Employing the techniques mentioned above, readers may add, edit, and reorder questions as desired. In choosing questions, readers should keep in mind how the data collected will be used and should limit themselves to questions that truly deliver the information they desire. We echo Radford's (2014) advice on assessment: "If you aren't able to do the analysis, don't collect the data." Needless questions contribute to respondents' fatigue and a sense of drudgery, emotions that do not need any further association with the library. A few short questions that get to the heart of the service provided and how to improve it should be the goal.

Once the relevant questions have been added and appropriately ordered, readers should view the live form to confirm that the details and layout are as desired. Readers also should attempt to submit a response in order to identify any problems before the form is widely deployed. Any trial responses submitted can easily be deleted by opening the associated spreadsheet and clearing the data. Trial responses should be deleted completely to avoid contaminating the official response pool.

As previously mentioned, when the form is deployed, its responses will automatically populate an associated Google sheet. This spreadsheet will have the same name as the form with "Responses" appended to its title and will be stored in the same folder as the form. Readers may choose to edit the destination of the responses for various reasons (e.g., when creating multiple forms whose responses all will be saved in a single spreadsheet). The destination can be altered by clicking the button Choose Response Destination. As similarly advised throughout this text, the response destination should not be changed once form responses have begun to be collected. Changing the response destination after responses have been received will scatter data among multiple spreadsheets and make analysis more difficult.

Feedback forms are simple to construct and provide a low-barrier channel for library users to share their constructive criticism. Following an instruction session, readers may log in to their Google Drive account, open the "Responses" spreadsheet, and identify the feedback responses by their timestamp and other identifying information. The responses received from feedback forms are typically short, anecdotal, and nonnumeric, so readers need not perform extensive analysis on the data in order to identify what went well and what could be improved.

CREATING A QUIZ

A more advanced application of Google Forms is to create a quiz, which readers might use to gauge students' retention of materials after an information-literacy session or workshop. There are several types of quizzes that can be created through Google Forms. The following example explains how to create a basic completion quiz that provides automatic feedback when respondents attempt to submit wrong answers. Like the podcast quiz at the University of Alabama, this format forces participants to answer all questions correctly and does not require time-consuming analysis of results. Additionally, unlike in the aforementioned feedback form example, questions in a quiz can be distributed across a series of pages, making the form a progressive series of questions.

To begin creating a quiz, readers may follow the steps laid out previously: creating and naming a new form, choosing a theme, and adding the first question. In order to route respondents to error pages, readers must use either the multiple-choice or choose-from-a-list question types. After creating the first question, readers should navigate to the Insert menu and click Page Break, which will create a new page after the first question. Readers will be prompted to assign a title to this new page, along with a description. Its title and description should inform respondents that they have answered incorrectly and may click the button below to try again. Following this error page, readers should insert a new page break.

Readers should continue in this pattern of inserting a new multiple-choice or list question followed by an error page. If a given question will have no right or wrong answer, readers should not include an error page after that particular question. For the quiz to work properly, all questions should be marked as required.

Once all the desired questions, page breaks, and error pages have been created, readers will set the quiz's navigational components by indicating the correct and incorrect answers. To start, readers should return to the first question. If the question type is multiple choice or choose from a list, readers should see a checkbox labeled "Go to page based on answer." Checking this box will display a drop-down menu beside each answer option. The menu allows readers to indicate where to send a respondent if that particular answer is chosen. For the correct answer, readers should change the drop-down menu so that it directs respondents to the next question (skipping the error page). For incorrect answers, the drop-down menu should be programmed to take respondents to the appropriate error page.

Readers will need to add one final navigational component to allow respondents to leave an error page and submit a new answer. Under each page in a Google form, readers will notice a drop-down menu, aligned to the right, that indicates where respondents will be directed after leaving that page. The default setting is to take respondents to the subsequent page. For each error page, this destination should be changed to navigate respondents back to the previous question, so that they may submit a new response.

As with the feedback form, readers may view quiz responses by logging in to their Google Drive account, opening the associated "Responses" spreadsheet, and identifying the feedback responses by their timestamp and other identifying information. Readers may use this data to identify how many students in a course section actually completed the quiz and plan accordingly. If few responses were submitted, for example, a reader might need to adjust his or her lesson plan.

Although this chapter has focused on introductory information for all levels of expertise, we recommend Google's support website (http://support. google.com/drive) for further reading on using Google Forms and on formulas for data manipulation in Google Sheets.

CONCLUSION

In the face of growing demands to demonstrate their value, librarians can use Google Forms to design intuitive, web-based forms for collecting patron feedback and for measuring student learning. Librarians may follow the examples given as to how the University of Alabama Libraries has incorporated Google Forms into its assessment strategy. Web-based assessment data is easier to implement, analyze, share, and store than paper-based questionnaires and assignments. Furthermore, Google Forms provides a cost-free alternative to the many proprietary programs available for library assessment. As the processes for collecting and using assessment data continue to change, librarians may find other, more complex applications of Google Forms for this purpose.

REFERENCES

Angell, Katelyn. 2013. "Open Source Assessment of Academic Library Patron Satisfaction." *Reference Services Review* 41 (4): 593–604.

Askew, Consuella A., and Eileen Theodore-Shusta. 2014. "How Do Librarians Learn Assessment?" *Library Leadership & Management* 28 (1): 1–9.

Radford, Marie L. 2014. "Getting the Right Fit: Tailoring Assessment Strategies for Your Library." Talk presented at the Online Computer Library Center event, Waltham, Massachusetts, April 22.

Whitlock, Brandy, and Julie Nanavati. 2013. "A Systematic Approach to Performative and Authentic Assessment." *Reference Services Review* 41 (1): 32–48.

Chapter Thirty

Using Google to Locate Government Information

Christopher C. Brown

While we don't need to tell anyone about the value of searching Google for information, focused searching for government information has some nuances that deserve special treatment. We are all accustomed to Google searches returning millions of results. But for the untrained searcher, we have no idea how deeply the results we really need are buried. Two problems exist: (1) We don't have a clear idea how Google ranks search results and (2) Even if we have all day, Google will not let us browse past the one thousandth search result (Harzing 2013). For these reasons we must employ some specialized search strategies to locate government information effectively. This chapter will cover U.S. federal government sites while also touching on U.S. state as well as foreign and international government information sites.

FINDING U.S. FEDERAL INFORMATION

The U.S. federal government is the largest publisher of information both in the print era and also now in the digital age. Some of the information disseminated resides in databases not visible to Google. This includes statistical information, massive amounts of scientific observations (e.g., NASA earth surface temperature data, NOAA National Climatic Data Center climate monitoring data, NOAA tides and currents data, USGS water data, USDA crop acreage data), economic data (e.g., Bureau of Economic Analysis, Treasury Department Data and Charts Center), and other data not of a textual nature. Indirect search strategies are necessary to uncover this information. For general references to websites, see appendix 30.1, which lists these sites by name and provides the URLs.

Open Government, Open Databases

Even though recent search engine optimization (SEO) technology has opened up much previously hidden database content, there are still massive amounts of government information completely invisible to Google. When we read about SEO, we often think about companies trying to place their results higher up within the organic search results of Google searches. White hat SEO is considered playing by the rules, whereas black hat SEO is considered cheating. However, a certain kind of SEO can be used not so much for positioning website data as exposing it in the first place. Structure optimization SEO can be used to expose database structures that previously have been invisible to Google crawls (Saberi, Saberi, and Mohd 2013). But for the vast amount of government data that remains invisible, searchers need to invoke an indirect search strategy: use Google to search for a database likely to contain the desired information, and then search the database itself for that information.

Most library catalogs are not crawled by Google, nor would we want them to be. Imagine a situation where online catalogs from all of the approximately 4,700 degree-granting institutions of higher education in the United States (National Center for Education Statistics 2012) were exposed to Google searching. A title search of a popular book might render results from many hundreds of libraries—not at all what we were looking for. Exposing of database structures to Google so that every catalog record can be crawled has no desirable effects in these cases. This illustrates the reality that having all data structures exposed to search engines is really not desirable.

However, in the case of some unique government databases, exposing the content is very beneficial. In their 2001 book, *The Invisible Web: Uncovering Information Sources Search Engines Can't See*, Sherman and Price (2001) note many databases that search engines cannot uncover for various reasons, including technological reasons (database structure issues that prevent crawling) and access reasons (behind a firewall that requires a password). Many of the databases they cite as not being accessible to Google are now crawled regularly. For whatever reason, whether it was changes in database structure, implementation of SEO structural optimization, or some other reason, many more government databases are now exposed to Google. However, there are still myriads of databases that are not. These tend to be nontextual databases such as those mentioned above.

Perhaps the best example of openness of access is the migration of Government Printing Office content from GPO Access to FDsys. Launched in 1994, GPO Access was the primary portal to recent congressional issuances, including bills, House and Senate reports and documents, the *Congressional Record*, and the *Congressional Directory*, among many other congressional sources. From the Executive Branch, GPO Access made searchable the *Budget of the United States, Federal Register, Code of Federal Regulations, Economic Indicators, Economic Report of the President,* and the *Weekly Compilation of Presidential Documents,* among many other sources. Each of the titles mentioned above were all in the "hidden Internet" category under the GPO Access interface—they couldn't be crawled by Google and couldn't be discovered by users.

The Federal Digital System (FDsys) was launched in 2009 after GPO Access content had been successfully migrated. The difference was that FDsys content was completely exposed to Google searching. This provided a major step in free public access to U.S. government information. Now every word of FDsys content, including PDFs, is indexed by Google.

Focusing Your Search with Top-Level Domains (TLDs)

Once we have identified likely government entities for our information needs, a focused Google search is in order. You already know what a top-level domain (TLD) is, although you may not have realized what it was. The familiar .com is a TLD, as are .edu, .gov, .mil, and dozens of others. Every country of the world has a TLD. Knowing this is important when it comes to searching Google, since, as previously mentioned, Google only lets users see the first one thousand results retrieved.

To do a site-specific or TLD Google search for a .gov domain, simply include *site:gov* in your Google search string. It is important not to put a space after the *site:*, as this will break the search syntax and produce incorrect results. A site-specific search retrieves Google's indexing of a domain or subdomain and gets around any search-engine-bias issues and exposes otherwise deeply buried results. Further, one could also specify secondary or tertiary subdomains in the site-specific search, such as *site:gpo.gov, site:bea.gov,* and *site:chroniclingamerica.loc.gov.*

This TLD strategy is also extremely useful when trying to find primary documents published by foreign governments. One can search Google using foreign languages, but it is amazing how many materials are released in English from governments the world over. To focus on foreign documents, first find a list of TLDs worldwide by doing a Google search for TLD (the *Wikipedia* page is usually toward the top of the list, and for these purposes this page is sufficient), and locate the country you are interested in. Then do a site-specific search within that country, something like this: *site:jp government.* Examine the results list to see if an obvious government subdomain appears. In the case of Japan, .go.jp is the government subdomain. You could then reframe your search like this: *site:go.jp <search terms>.*

It should be noted that not all foreign governments have the same governmental secondary domain. Searchers need to pay close attention to local

practices. Table 30.1 displays some of the variations in secondary domains
across selected countries.

Now back to the United States. Site-specific searching is an extremely
effective method for U.S. government entities. However, U.S. government
information is hosted on nearly two thousand domains. For example, the
Louisiana State University Federal Agency Directory contains nearly two
thousand lines with links to government offices. In 2004, the University of
Denver did an examination of URLs in their local catalog and discovered that
there were about 1,400 distinct domains represented at that time (Brown
2004). It is important to note that although most government domains have a
.gov TLD, not all do. Here are some notable exceptions among governmental
and quasi-governmental domains.

- United States Postal Service prefers to use usps.com ; usps.gov flips over
  to usps.com
- United States Army War College Strategic Studies Institute is an example
  of a military domain, http://www.strategicstudiesinstitute.army.mil
- United States Forest Service, under the Department of Agriculture
  (usda.gov), uses www.fs.fed.us

Table 30.1. Selected Secondary Domains for Foreign Governments

Brazil	gov.br
Canada	gc.ca
China	gov.cn
Croatia	vlada.hr
Ecuador	gob.ec
France	gouv.fr
Japan	go.jp
Mexico	goub.mx
Peru	gob.pe
Russia	gov.ru
Senegal	gouv.sn
South Africa	gov.za
South Korea	go.kr
Sweden	government.se
Switzerland	gov.ch
Venezuela	gob.ve
Zimbabwe	gov.zw

- FRASER—Federal Reserve Archive from the Federal Reserve Bank of St. Louis hosts its service at fraser.stlouisfed.org

It doesn't make sense, however, to do site-specific searching for every government site. For example, Regulations.gov is indexed by Google, but Google retrieves results for issues that are no longer available for comment. Using the Regulations.gov website directly avoids this problem. Science Accelerator is not indexed by Google, thus, there is no point searching it. Congress.gov is crawled by Google, but the results presented in the native interface far outweigh anything Google can provide.

Understanding Government Structure

While it is possible to just start searching Google directly, it is most helpful to have an outline of the federal government together with their Internet addresses in mind, so that focused searches can be done. Having this knowledge of specific government websites enables us to do site-specific searches more effectively.

Partnering with the Government Printing Office (GPO), Louisiana State University Libraries has created a Federal Agency Directory (LSU Libraries 2014) that helps users understand the many government agencies, boards, commissions, and committees, both official and quasi-official. Those wanting a more official take on government structure should consult the *United States Government Manual* (U.S. Office of the Federal Register 2013).

Google Books and Government Documents

Since U.S. government documents are generally not copyrighted, they are in the public domain. Digital scanning projects such as Google Books have many of these documents available for full view. Be aware that in some cases Google Books may not have released a government document to full view. In these cases it is worth the time to check the same title in the HathiTrust or Internet Archive's Digital Book Collections to see if full access is available. HathiTrust estimates that approximately 4 percent of its content is U.S. federal publications (HathiTrust 2013), all of which are viewable through that program but may not be viewable in Google Books.

Locating Government-Issued Statistics with Google

Finding statistics is difficult for several reasons. First of all, they are numbers; we can't really search for numbers in the same way we can search for textual information. Second, many statistics are buried in government databases that may not be open to Google's crawling. We need to invoke different strategies when searching Google. When trying to find textual information, try searching Google in one of two ways: search for the anticipated "name" or title of the answer, either in the form of a statement or in the form of a question (e.g., "unemployment rate in Colorado"), or search for information we expect to find in the answer (e.g., "2013 enplanements Denver"). Perhaps the most effective strategy when searching for statistics is an indirect one: search for the government entity likely to publish the information.

The most important question to ask is, what government agency is most likely to publish the statistics I am looking for? For this we can look to FedStats.gov . Operated by the Census Bureau, FedStats provides links to over one hundred federal agencies that issue statistics. Another source to check is the *Statistical Abstract of the United States* (U.S. Census Bureau 2011). Even though this Census Bureau publication is no longer published, it nevertheless provides many clues as to which federal agencies publish what statistics.

Searching Government Portals through the Google Back Door

Selected government information portals can be searched both through the portal's own native interface as well as via Google. It's not that searching Google in these cases is better; it's that the indexing of results is different and may work better in individual cases. FDsys, previously mentioned, is a good example of a search portal exposed to Google. Searching for State of the Union messages, for example, is quite effective using Google (*site:gpo.gov fdsys "state of the union"*). Most of the time, however, it is best to search the native FDsys interface, which, with its "next-gen" interface, is easy to use, with facets or limiters along the left side for easy limiting of results.

Other federal sites that work well using a Google site-search strategy include the National Center for Education Statistics (e.g., *site:nces.ed.gov 4 year colleges tuition*), the National Climate Data Center (e.g., *site:ncdc.noaa.gov denver precipitation*), and the Bureau of Labor Statistics (e.g., *site:bls.gov unemployment*).

One caution to be aware of is that Google is not able to distinguish superseded information from current information. For example, the website earmarks.omb.gov seems to be current only through mid-2011. It is no longer maintained but is still live and is indexed in Google.

FINDING STATE GOVERNMENT INFORMATION

Previously the .gov domain was just for the federal government. But now state and local entities may also have .gov TLDs (Federal Management Regulation 2003). Originally the .us domain was designed for states to use, following this pattern: state.xx.us—where xx is the two-letter state postal abbreviation. But most states decided that they wanted better branding, so they added domains they deemed easier to remember, such as myflorida.gov and ny.gov. Table 30.2 shows some of these variations. It is important when searching for state government information to search each of these domains separately, as content is often distributed over different Internet domains.

OLDER DOCUMENTS—WHAT TO DO WHEN GOOGLE DOESN'T COME THROUGH

While current government information may be relatively easy to locate using the Google search strategies mentioned so far, agencies do not keep information on the web forever. That job is usually the task of libraries and archives. They see their mission as archiving older information for all time. Yet there are some government entities that are interested in providing access to older documents, such as FRASER from the Federal Research Bank of St. Louis (FRASER 2014), the Library of Congress's American Memory collections (Library of Congress 2014), and the U.S. Department of Labor Digital Library. But generally, government publications tend to disappear from government websites. Several initiatives exist to mitigate this problem.

Government Printing Office Persistent URLs (PURLS)

The Government Printing Office, when it catalogs online government content, creates persistent URLs, or PURLS, and at the same time it archives the digital content in case it disappears from the agency site (Federal Depository Library Program 2010). Mechanisms are also in place for documents librarians to report "fugitive" online documents to GPO so that they can be cataloged and archived in digital format. Documents cataloged by GPO are searchable in the Catalog of U.S. Government Publications (CGP). CGP records are not exposed through Google, so it may be necessary to search for PURLS using the CGP itself.

Internet Archive's Wayback Machine

Sometimes the Wayback Machine is useful in retrieving lost documents. The 2001 title *Babies Sleep Safest on Their Backs: Reduce the Risk of Sudden Infant Death Syndrome (SIDS)*, was once online at http://www.nichd.nih.gov/sids/sleep_risk.htm , but disappeared long ago. A search of the Wayback Machine will enable users to retrieve the content (see Back to Sleep Campaign 2001).

Table 30.2.   Domains for U.S. States

State	Domains with Site-searchable Content
Alabama	al.gov redirects to alabama.gov; content also under state.al.us
Alaska	alaska.gov; content also under state.ak.us
American Samoa	as.gov and americansamoa.gov
Arizona	arizona.gov redirects to az.gov and arizona.gov; also uses azleg.gov; azcourts.gov; and other .gov domains; content also under state.az.us
Arkansas	ar.gov and arkansas.gov; content also under state.ar.us
California	ca.gov; content also under state.ca.us
Colorado	co.gov redirects to colorado.gov; content also under state.co.us
Connecticut	ct.gov; content also under state.ct.us
Delaware	de.gov and delaware.gov; content also under state.de.us
District of Columbia	dc.gov; content also under dc.us
Florida	florida.gov and fl.gov both redirect to myflorida.com, but much content under both .gov domains; content also under state.fl.us
Georgia	georgia.gov; content also under ga.gov and state.ga.us
Guam	guam.gov
Hawaii	hawaii.gov redirects to ehawaii.gov, but content under both domains; content also under state.hi.us
Idaho	idaho.gov; content also under state.id.us
Illinois	il.gov redirects to illinois.gov, but content under both domains; content also under state.il.us
Indiana	in.gov and indiana.gov; content also under state.in.us
Iowa	ia.gov and iowa.gov; content also under state.ia.us
Kansas	ks.gov and kansas.gov; content also under state.ks.us

Kentucky	ky.gov redirects to kentucky.gov, but content under both domains; content also under state.ky.us
Louisiana	la.gov redirects to louisiana.gov; but content under both domains; content also under state.la.us
Maine	maine.gov; content also under state.me.us
Maryland	md.gov redirects to maryland.gov; but content under both domains; content also under state.md.us
Massachusetts	ma.gov and massachusetts.gov both redirect to mass.gov, but all three domains have content; content also under state.ma.us
Michigan	mi.gov and michigan.gov; content also under state.mi.us
Minnesota	minnesota.gov redirects to mn.gov, but content under both domains; content also under state.mn.us
Mississippi	mississippi.gov redirects to ms.gov, but content under both domains; content also under state.ms.us
Missouri	missouri.gov redirects to mo.gov, but content under both domains; no content under state.mo.us
Montana	montana.gov and mt.gov, with content under both domains; content also under state.mt.us
Nebraska	ne.gov and nebraska.gov, with content under both domains; content also under state.ne.us
Nevada	nevada.gov redirects to nv.gov, but content only under nv.gov; content also under state.nv.us
New Hampshire	nh.gov; content also under state.nh.us
New Jersey	nj.gov and newjersey.gov; content also under state.nj.us
New Mexico	newmexico.gov; content also under state.nm.us
New York	ny.gov; content also under state.ny.us
North Carolina	nc.gov; content also under state.nc.us

North Dakota	northdakota.gov redirects to nd.gov, but content only under nd.gov; content also under state.nd.us
Ohio	oh.gov redirects to ohio.gov; local content under oh.gov, state content under ohio.gov; content also under state.oh.us
Oklahoma	ok.gov and oklahoma.gov, most content under ok.gov; content also under state.ok.us
Oregon	oregon.gov; most content under state.or.us
Pennsylvania	pennsylvania.gov redirects to pa.gov, most content under pa.gov; content also under state.pa.us
Puerto Rico	puertorico.gov redirects to pr.gov, most content under pr.gov
Rhode Island	rhodeisland.gov redirects to ri.gov, content only under ri.gov; content also under state.ri.us
South Carolina	sc.gov; content also under state.sc.us
South Dakota	sd.gov; content also under state.sd.us
Tennessee	tn.gov and tennessee.gov; content also under state.tn.us
Texas	tx.gov redirects to texas.gov, but content under both domains; content also under state.tx.us
Utah	utah.gov; content also under state.ut.us
Vermont	vt.gov and vermont.gov; content also under state.vt.us
Virginia	virginia.gov; content also under state.va.us
Virgin Islands	vi.gov
Washington	washington.gov redirects to wa.gov, but content only under wa.gov; content also under state.wa.us
West Virginia	wv.gov; content also under state.wv.us
Wisconsin	wi.gov and wisconsin.gov; content also under state.wi.us
Wyoming	wy.gov and wyoming.gov; content also under state.wy.us

End of Term Web Archive

The End of Term Web Archive, begun in 2008, is a collaborative endeavor between the California Digital Library, the Internet Archive, the Library of Congress, University of North Texas Libraries, and the Government Printing Office. Since government websites are in a state of transition between presidential terms, the End of Term Web Archive provides a snapshot of selected government web content during a transition period. This project is mentioned at this point not because it is searched via Google, but because this government content would not be discovered using the Google search strategies mentioned in this chapter.

Federal Depository Library Program

Websites are not libraries. Their purpose is not to archive all documents for all time. Thus, Google will often be unable to retrieve older documents. This is the role of libraries, and in the case of the United States, we have a program for that. It's the Federal Depository Library Program (FDLP), administered through the Government Printing Office. At this point, indirect Google search strategies come into play. One must first locate a depository library near them using GPO's online library directory (Federal Depository Library Program 2012). Then searching that library's online catalog will reveal older cataloged publications.

Researchers need to realize the added value that libraries bring to the search for government information. Not only do they have the resources traditionally gathered in by the FDLP in tangible formats and recently in online formats, but also the many licensed resources that cover government information. This may require searching a library's online catalog and searching individual licensed information portals containing government information.

CONCLUSION

Governments the world over are increasingly preferring the web as their publishing venue. Knowing how to bring this information to the surface with site-specific searching, as well as when and how to search for hidden-Internet content, is important in the quest for these primary sources. Realizing what to do when Google does not come through will help librarians and researchers find government information more expeditiously.

APPENDIX 30.1

Bureau of Economic Analysis—http://www.bea.gov/

Catalog of U.S. Government Publications (CGP)—http://catalog.gpo.gov/

End of Term Web Archive—http://eotarchive.cdlib.org/

FRASER, from the Federal Research Bank of St. Louis—http://fraser.
stlouisfed.org/

HathiTrust—http://www.hathitrust.org/

Internet Archive's Digital Books Collections—https://archive.org/details/
texts

NASA earth surface temperature data—http://data.giss.nasa.gov/gistemp/

NOAA National Climatic Data Center climate monitoring data—http://
www.ncdc.noaa.gov/

NOAA tides and currents data—http://tidesandcurrents.noaa.gov/

U.S. Department of Labor Digital Library—http://www.dol.gov/oasam/
wirtzlaborlibrary/digital/

U.S. Department of the Treasury Data and Charts Center—http://www.
treasury.gov/resource-center/data-chart-center/Pages/index.aspx

USGS Water Data—http://waterdata.usgs.gov/nwis

USDA crop acreage data—http://www.fsa.usda.gov/FSA/webapp?area=
newsroom&subject=landing&topic=foi-er-fri-cad

REFERENCES

Back to Sleep Campaign. 2001. *Babies Sleep Safest on Their Backs: Reduce the Risk of Sudden Infant Death Syndrome (SIDS)*. Bethesda, MD: Maternal and Child Health Bureau, U.S. Department of Health and Human Services National Institutes of Health. Content recovered from the Wayback Machine, http://web.archive.org/web/20050404151717/http://www.nichd.nih.gov/sids/sleep_risk.htm .

Brown, Christopher C. 2004. "Knowing Where They're Going: Statistics for Online Government Document Access through the OPAC." *Online Information Review* 28 (6): 396–409.

Federal Depository Library Program. 2010. "Linking to Federal Resources Using Persistent Uniform Resource Locators (PURLs)." Updated February 21, 2014. http://www.fdlp.gov/requirements-guidance/instructions/709-purls .

———. 2012. "Federal Depository Libraries" [Library Directory]. Updated March 24, 2014. http://www.fdlp.gov/about-the-fdlp/federal-depository-libraries .

Federal Management Regulation. 2003. Internet GOV Domain, 68 *Fed. Reg.* 15089 (March 28).

FRASER, Federal Reserve Archival System for Economic Research. 2014. St. Louis: Federal Reserve Bank of St. Louis. http://fraser.stlouisfed.org/.

Harzing, Anne-Wil. 2013. "A Preliminary Test of Google Scholar as a Source for Citation Data: A Longitudinal Study of Nobel Prize Winners." *Scientometrics* 94 (3): 1057–75.

HathiTrust. 2013. "What Is HathiTrust?" http://www.hathitrust.org/documents/HathiTrust-Overview-Handout.pdf .

Library of Congress. 2014. *The Library of Congress American Memory*. Washington, DC: Library of Congress. http://memory.loc.gov/ .

Louisiana State University (LSU) Libraries. 2014. *Federal Agency Directory*. http://www.lib.lsu.edu/gov/ .

National Center for Education Statistics. 2012. *Digest of Education Statistics*. Washington, DC: U.S. Department of Health, Education, and Welfare, Education Division, National Center for Education Statistics. Table 306.

Saberi, Saeid, Golnaz Saberi, and Masnizah Mohd. 2013. "What Does the Future of Search Engine Optimization Hold?" *International Journal of New Computer Architectures and Their Applications* 3 (4): 132–38.

Sherman, Chris, and Gary Price. 2001. *The Invisible Web: Uncovering Information Sources Search Engines Can't See*. Medford, NJ: Information Today.

U.S. Census Bureau. 2011. *Statistical Abstract of the United States, 2012*. Washington, DC: Government Printing Office. https://www.census.gov/compendia/statab/ .

U.S. Office of the Federal Register. 2013. *United States Government Manual*. Washington, DC: Office of the Federal Register, National Archives and Records Service, General Services Administration. http://www.gpo.gov/fdsys/browse/collection.action?collectionCode= GOVMAN .

Index

About the Contributors

Jaena Alabi is a reference librarian at Auburn University's Ralph Brown Draughon Library, Auburn, Alabama, where she serves as liaison to the English and psychology departments. She earned an MA in English in 2005 and an MLIS in 2006, both from the University of Alabama. Jaena is a member of the Association of College and Research Libraries and has presented at several ACRL National Conferences. She has collaborated with her coauthor, William H. Weare Jr. (Indiana University–Purdue University Indianapolis), on two conference presentations, a conference paper, and a peer-reviewed article.

Seth Allen is the outreach librarian for the Adult and Graduate Studies (AGS) division of Bryan College, Dayton, Tennessee. He obtained his MLIS degree from the University of North Carolina at Greensboro and is currently pursuing an MA in new media and global education. Seth has worked in academic and public libraries for nearly eight years and presented at several education and library conferences on the use of new media in instruction. Seth recently published "Towards a Conceptual Model of Academic Libraries' Role in Student Retention" in the spring 2014 issue of the *Christian Librarian.*

Laura Baker is the librarian for digital research and learning at Abilene Christian University in Abilene, Texas. She explores how digital information changes the way people research and how those changes affect library services. Her professional interests include emerging technologies, holistic user experience, and open source solutions. She has given presentations on library learning commons at ALA and LOEX and most recently authored an article about QR codes for the *Journal of Library Innovation*. Laura holds an MLIS from the University of Texas at Austin and a BBA in computer information systems from Abilene Christian University.

Theresa Beaulieu is the education librarian at the University of Wisconsin–Milwaukee. She earned her MLIS from the University of Arizona and her MEd from George Washington University in the District of Columbia. Theresa worked as a program manager for the College of Education at Arizona State University, was the director of education and cultural affairs for the Stockbridge-Munsee Community in Wisconsin, and taught children's literature to prospective educators. She is interested in improving access and equity, enjoying children's literature with others, and creating user-centered spaces.

Heather Beirne is an education librarian at Eastern Kentucky University Libraries. She holds an MSLS from the University of Kentucky and a bachelor's in English from Transylvania University in Lexington, Kentucky. Heather has held positions in both public and academic libraries throughout central Kentucky. She has published in *Kentucky Libraries* and presented at conferences for the Kentucky Library Association and at many other state and regional conferences related to librarianship and education. Her research interests include information literacy, web 2.0, digital literacy, digital citizenship, and digital storytelling.

Laura Bohuski is a database project specialist at Western Kentucky University (WKU) where she fixes human and computer errors that develop in the library's ex-libris database. She received her bachelor's in history and social studies from WKU before receiving her master's in library science from Indiana University Bloomington in December 2010. Laura has started work on her master's in history from WKU, is a new member of the American Library Association, and looks forward to interacting with the academic community.

Robbie Bolton is the library director at the White Library at Spring Arbor University in Spring Arbor, Michigan. He received an MS in information from the University of Michigan. He serves on the editorial board for the *Journal of the Association of Christian Librarians: The Christian Librarian.* He is also a community representative for the Digital Public Library of America. An avid runner, cyclist, and hiker, Robbie also teaches a backpacking class every semester and enjoys hiking on the Appalachian Trail. He resides in Jackson, Michigan, with his wife and three children.

Christopher C. Brown is reference technology integration librarian/government documents librarian at the University of Denver, University Libraries.

He enjoys providing research guidance to students, especially in matters of international studies, public policy, and government information. His passions include providing discovery tools that allow users to find the rich resources to which the library subscribes and providing free public access to government publications. As government documents librarian, he has served on the Depository Library Council to the United States Public Printer. He also teaches in the university's Library and Information Science program.

Sarah Cohn is a visiting assistant librarian of information services at Manhattan College, Bronx, New York. She is a pastry chef turned librarian, historical mystery fan, and triathlete. She has an MLIS from the City University of New York Queens College in Queens, New York, and is a fiction reviewer for *Library Journal*. She has presented at the Roger Smith Cookbook Conference on cookbooks as works of art and is a volunteer instructor of information literacy at Legal Outreach's College Bound program in Queens, New York.

Ashley Cole is a reference and instruction librarian at Eastern Kentucky University, specializing in student engagement and the first-year experience. In this role, Ashley collaborates with formal and informal undergraduate campus communities to develop, implement, and assess services and programs aimed at improving student success. She has conducted workshops at the Kentucky Library Association, Pedagogicon, and the National Personal Librarian and First Year Experience Library Conference. Along with a bachelor's from Eastern Kentucky University, she also holds a master's degree in library and information science from the University of Kentucky.

Amanda Dinscore is a public services librarian at California State University (CSU), Fresno. She earned her MLIS from San Jose State in 2007 and a certificate in online teaching and learning from CSU East Bay in 2013. Her interests include website usability, instructional design, and online learning. She has presented at state and national conferences on web usability and recently published an article with her colleagues in *Internet Reference Services Quarterly*. She is a 2010 ALA Emerging Leader and regularly conducts workshops for students on using Google tools for academic research.

Chelsea Dodd is the reference and adult programming supervisor of Montclair Public Library in Montclair, New Jersey. With a background in marketing communications, her career goal is to help bring energy back to the public library by employing modern technologies and expanding community outreach. Chelsea earned her MLIS from Drexel University in Philadelphia, and BA in communication from Seton Hall University in South Orange, New

Jersey. Her favorite things to Google include recipes and travel itineraries and she blogs at librarianchels.wordpress.com.

Wei Fang, a tenured library faculty at Rutgers University in New Jersey, has used Google Analytics and other web analytic utilities extensively since 2006. He was a co-panelist in ALA Midwinter Conference's panel in 2010, titled "Analytics and Statistics—Improving Library Services through Google Analytics and Other Tools." Wei has published peer-reviewed journal articles on Google Analytics and also on video repositories. He published the first article on using Google Analytics in libraries in *Library Philosophy and Practice*, titled "Using Google Analytics for Improving Library Website Content and Design: A Case Study."

Adam Fullerton is the current library director at Morningside College's Hickman-Johnson-Furrow Library in Sioux City, Iowa. He received an MLIS from the School of Information Sciences at the University of Wisconsin–Milwaukee in 2010 after completing a BA in history from the University of Wisconsin–Superior in 2009. Adam is a member of ALA, ACRL, Iowa Library Association, and Iowa ACRL. In his free time, he volunteers for a number of local nonprofit organizations, including the Sioux City Historic Preservation Commission and Sioux City MidAmerican Museum of Aviation and Transportation.

Teri Oaks Gallaway is the library systems and web coordinator at Loyola University New Orleans. In addition to her work on user-experience studies, she participates in the library's reference and instruction programs as the liaison to the sociology, criminal justice, and honors departments. She has published and presented on topics including selection of a web-scale discovery system, promoting electronic resources, and using web services APIs to display data from library systems. She has used Google Drive as an assessment tool for the evaluation of serials subscriptions, website traffic, and circulation statistics.

Neyda V. Gilman is a learning commons librarian at Syracuse University's Bird Library in Syracuse, New York. She has a BS in medical laboratory science from the University of Utah, and her first career was working as a medical technologist in a variety of laboratories. She earned her MLIS from the University at Buffalo to begin her second career as a librarian. Currently she is an officer in the Biomedical and Life Sciences Division of the Special Library Association as well as a member of the Medical Library Association.

Samantha Godbey, education librarian at the University of Nevada, Las Vegas, received her MLIS from San Jose State University and MA in education from the University of California at Berkeley. Prior to becoming a librarian, Samantha was a high school English teacher in Berkeley, California, and served as a Peace Corps volunteer in the Russian Far East. Her research focuses on collaboration and information literacy instruction. She presented on her experience with embedded librarianship at the Library 2011 Worldwide Virtual Conference.

Deborah Hamilton is the learning technologies librarian for Stitt Library at Austin Presbyterian Theological Seminary in Austin, Texas. In addition to her library duties, Deborah works with faculty on the development and production of online courses. Deborah earned her master of arts in English language and literature from the University of Maryland in 2004, her master of education from Antioch University Seattle in 2006; and her master of library and information science from the University of Washington in 2011. Her professional memberships include the American Library Association, Access, and Educause.

Amy Handfield is assistant librarian for access services at Manhattan College, Bronx, New York. She a visual artist and librarian who appreciates the network formed by her library colleagues. She has an MFA from Goddard College in Plainfield, Vermont, and an MLS from Drexel University in Philadelphia, Pennsylvania. Her research interests include technology applications in library administration and the "bring your own device" pedagogical movement. She has presented her research at the CUNY EdTech Showcase and Georgia Tech's Access Services conference.

Amy James is an emerging technologies librarian at the White Library at Spring Arbor University in Spring Arbor, Michigan. She obtained her MLIS from Wayne State University with a concentration in academic librarianship. Amy's memberships include the American Library Association, Michigan Library Association, Library Instruction Round Table, and the Association of Christian Librarians. She enjoys researching library technology, user-centered web design, and the history and developments of academic libraries in China. She resides in Jackson, Michigan, with her husband, and they enjoy traveling, kayaking, and being outdoors.

Alejandro Marquez, reference and instruction librarian at Reed Library on the Fort Lewis College Campus, Durango, Colorado, obtained his MLIS from the University of Denver. Before his current position, he served in the Peace Corps El Salvador as municipal development volunteer. Alejandro's memberships include Reforma Colorado and the American Library Association (ALA). He has appeared in ALA's *Library Worklife* and blogs on the

Colorado State Library blog. His research interests include library services for minority populations and the use of technology in libraries.

Rafia Mirza is the digital humanities librarian and liaison to women's studies at the University of Texas at Arlington Libraries in Arlington, Texas. She completed a master of science in information (MSI) at the University of Michigan, Ann Arbor, and a BA in psychology, American studies, and English at the State University of New York at Buffalo. Her research interests include technologies promoting information literacy instruction, public perceptions of Generation X librarians, comics, social media, open access, and digital humanities.

Apryl C. Price, electronic resources collection management librarian at Florida State University Libraries, Tallahassee, Florida, obtained her MLIS from the University of South Florida. Apryl's memberships include the American Library Association (ALA), the Association for College and Research Libraries (ACRL), the Florida Library Association (FLA), and the Black Caucus of the American Library Association (BCALA). Her writing has appeared in *Collection Building, Journal of Electronic Resources Librarianship, Serials Librarian, Library Collections, Acquisitions, & Technical Services*, and *Handbook of Research on Computer Mediated Communication*.

Sarah Richardson is the business librarian and team leader at the Business Library and Academic Commons at Eastern Kentucky University Libraries in Richmond, Kentucky. She obtained her master's in library and information science from the University of Kentucky. She also holds a bachelor's in psychology from Radford University in Radford, Virginia. During her library career, Sarah has held positions in several academic libraries in Kentucky and Virginia. She is a member of the Kentucky Library Association and BRASS, the Business Reference and Services Section of ALA's Reference and User Services Association (RUSA).

Steven Richardson is a reference librarian and assistant professor in the Law Library at Indiana Tech Law School. He teaches Legal Research and Citation as part of the Lawyering Skills team. Steven received his bachelor of arts degrees from the University of Kentucky as well as his juris doctor and his master's in library science. Steven is a member of the bar associations of Kentucky and Missouri and is a former member of the American Immigration Lawyers Association. His areas of practice were family reunification and humanitarian issues in immigration and dependency, neglect, and abuse cases as a guardian *ad litem*/court appointed counsel in the Family Court of Fayette County, Kentucky.

Mark Daniel Robison, research librarian at Valparaiso University, in Valparaiso, Indiana, obtained his master of library science from Indiana University Bloomington. Mark's memberships include the Alabama Library Association, the Association of College and Research Libraries, and the Library Instruction Round Table, American Library Association. Mark serves on LIRT's Transitions to College Committee. He has contributed reviews to the Internet Resources column in *Public Services Quarterly* and presented at the Alabama Library Association 2013 Annual Convention. Mark volunteers with the Kentuck Art Center in Northport, Alabama.

Josh Sahib, distance education librarian at the University of Alabama, Tuscaloosa, Alabama, received both his MLIS and his MA in American studies from the University of Alabama. His professional areas of focus include instruction, outreach, and emerging technologies. Josh participated in the Minnesota Institute for Early Career Librarians in 2012. He coauthored an article on 3D printing that appeared in *Issues in Science and Technology Librarianship*. He has presented at the UA Scholars Institute, the Alabama Library Association Annual Convention, and the Mississippi State eResource and Emerging Technologies Summit.

Corey Seeman, director of the Kresge Library of the Ross School of Business, University of Michigan in Ann Arbor, has been the director of Kresge Library since 2006. He previously was associate dean for resource and systems management at the University of Toledo (Ohio); library training consultant, Innovative Interfaces, Inc.; and manager of technical services, National Baseball Library at the National Baseball Hall of Fame and Museum, Inc., in Cooperstown, New York. He has an MALS (1992) from Dominican University and an AB (1986) from the University of Chicago. His popular *A Library Writer's Blog* is an important professional resource.

Carol Smallwood received an MLS from Western Michigan University and an MA in history from Eastern Michigan University. *Librarians as Community Partners: An Outreach Handbook* and *Bringing the Arts into the Library* are recent ALA anthologies. Others are *Women on Poetry: Writing, Revising, Publishing and Teaching* (2012); *Marketing Your Library* (2012); and *Library Services for Multicultural Patrons: Strategies to Encourage Library Use* (2013). Her library experience includes school, public, academic, and special libraries, as well as administration and being a consultant; she's a multiple Pushcart nominee.

Misti Smith, emerging technologies librarian at Mount Aloysius College, Cresson, Pennsylvania, obtained her MLS from Clarion University in Clarion, Pennsylvania. Recent successes include administration of a grant-funded iPad program and an accepted proposal for a technology commons in the library. She will appear in an upcoming issue of *LOEX Quarterly* and is a *Choice Reviews* contributor. Misti has recently presented workshops at the Virginia Library Association Annual Conference and was a keynote speaker at the Mid-Atlantic Library Alliance fall meeting. Misti is a member of the Ebensburg Women's Club and a local book club.

Aline Soules, library faculty member at California State University, East Bay, has an MSLS (Wayne State University), an MA (University of Windsor), and an MFA in creative writing (Antioch University Los Angeles). Her recent work has appeared in the journals *New Library World* and *Journal of Electronic Resources Management*, and she has book chapters in *Library Reference Services and Information Literacy: Models for Academic Institutions* (2013) and *Continuing Education for Librarians: Essays on Career Improvement through Classes, Workshops, Conferences and More* (2013).

Jennifer Starkey is a reference and instruction librarian at Oberlin College Library in Oberlin, Ohio, where she oversees reference services, instructs students in doing research in social sciences and humanities disciplines, and is liaison to the history and economics departments. She has presented on topics including assessment of research consultation programs and working with faculty to design effective research assignments. She works in a team-managed department that relies heavily on Google Drive for daily communication.

Fantasia A. Thorne-Ortiz has been employed at Syracuse University Bird Library, in Syracuse, New York, since 2009 as a learning commons librarian and also at the Onondaga County Public Library as a substitute part-time librarian. She holds a BA in English from Southern Connecticut State University and an MLIS from Simmons College. She is a member of the American Library Association and currently serves as the chair of the Office of Literacy and Outreach Services Advisory Committee, as well as a member of the Black Caucus of the American Library Association.

Sarah Troy, head of user services and resource sharing at the University of California, Santa Cruz, obtained her bachelor's degree in modern literature from the University of California, Santa Cruz, and her MLIS from San José State University. She has published in the *Journal of Access Services* and contributed to *Letting Go of Legacy Services: Case Studies* (2014). She currently oversees five public-service units, including circulation, collection maintenance, interlibrary loan, the media center, and reserves. Sarah is interested in issues related to leadership and management, resource sharing, and public service.

Diana K. Wakimoto, online literacy librarian at California State University, East Bay University Libraries, Hayward, California, obtained her MSLIS from Simmons College and her PhD from Queensland University of Technology via the San José State University Gateway PhD program. Diana's memberships include the American Library Association, Association of College and Research Libraries, and Society of American Archivists. Her work has been published in *The American Archivist, Archival Science, Evidence Based Library and Information Practice*, and *Library Hi-Tech*. Diana has presented at national and international conferences. She blogs at *The Waki Librarian*.

William H. Weare Jr. is the access services team leader at the University Library at Indiana University–Purdue University Indianapolis. He earned his MA in library and information science from the University of Iowa in 2004. William is a member of the Association of College and Research Libraries and the Library Leadership and Management Association. His work has appeared in *Collaborative Librarianship, Journal of Access Services, Library Leadership & Management*, and *Qualitative and Quantitative Methods in Libraries*. William and his coauthor for this book, Jaena Alabi (Auburn University), have collaborated on several conference presentations and several publications.

Andrew Weiss, digital services librarian at California State University, Northridge, in Los Angeles's San Fernando Valley is a graduate of the University of Hawai'i at Manoa (MLIS), Temple University (MA English), and Kenyon College (BA English). His research interests include massive digital libraries, institutional repositories and open access, and the history of Japanese libraries. His work has appeared in *OCLC Systems and Services, Journal of Library Metadata, Digitization in the Real World*, and many other publications; he authored *Using Massive Digital Libraries: A LITA Guide* (2014).

Peace Ossom Williamson is an informationist and works as liaison librarian at University of Texas at Arlington Libraries in Arlington, Texas. She received an MLS and an MS in health studies from Texas Woman's University and is a member of the Academy of Health Information Professionals. Peace has appeared in *Virtual Reference in Modern Libraries* (forthcoming), presented for regional and joint Medical Library Association meetings. She has also received numerous awards, including the Association of Research Libraries Initiative to Recruit a Diverse Workforce fellowship.

George L. Wrenn is the cataloging librarian and electronic resources coordinator at Humboldt State University, Arcata, California. He obtained his MLIS from the University of California, Los Angeles. He has twenty-five years of cataloging experience and has co-managed his university's institutional repository, Humboldt Digital Scholar, since 2005. His recent scholarship has focused on electronic access to scholarly materials. He has written about Humboldt Digital Scholar for *D-Lib Magazine*, and his article "Hidden in Plain Sight? Records for On-Demand Academic Public Lectures in OCLC WorldCat: A Survey" appeared in *Cataloging & Classification Quarterly*.